Promoting the National Spatial Data Infrastructure Through Partnerships

Mapping Science Committee
Board on Earth Sciences and Resources
Commission on Geosciences, Environment, and Resources
National Research Council

NATIONAL ACADEMY PRESS
Washington, D.C. 1994

NOTICE: The project that is the subject of this report was approved by the Governing Board of the National Research Council, whose members are drawn from the councils of the National Academy of Sciences, the National Academy of Engineering, and the Institute of Medicine. The members of the committee responsible for the report were chosen for their special competences and with regard for appropriate balance.

This report has been reviewed by a group other than the authors according to procedures approved by a Report Review Committee consisting of members of the National Academy of Sciences, the National Academy of Engineering, and the Institute of Medicine.

Support for this study by the Mapping Science Committee was provided by the Defense Mapping Agency, the United States Geological Survey, the Bureau of Land Management, and the Bureau of the Census.

Library of Congress Catalog Card No. 94-66772
International Standard Book Number 0-309-05141-X

Additional copies of this report are available from

National Academy Press
2101 Constitution Avenue
Box 285
Washington, D.C. 20418
800-624-6242
202-334-3313 (in Washington Metropolitan Area)

B-499

COVER: An early North American partnership—an etching of the marriage of Pocahontas and John Rolfe at Jamestown, Virginia, April 5, 1614. SOURCE: Library, Richmond (VA) Newspapers, Inc.

Copyright 1994 by the National Academy of Sciences. All rights reserved.

Printed in the United States of America

MAPPING SCIENCE COMMITTEE

LARRY J. SUGARBAKER, Washington State Department of Natural Resources, Olympia, *Chairman*
LAWRENCE F. AYERS, Intergraph Corporation, Reston, VA, *Vice-Chairman*
HUGH N. ARCHER, PlanGraphics, Inc., Frankfort, KY
WILLIAM M. BROWN, Environmental Research Institute of Michigan, Ann Arbor
BARBARA P. BUTTENFIELD, State University of New York, Buffalo
MICHAEL W. DOBSON, Rand McNally and Company, Skokie, IL
FREDERICK J. DOYLE, McLean, VA (retired, U.S. Geological Survey)
MICHAEL J. FOLK, University of Illinois, Urbana
LEE C. GERHARD, Kansas Geological Survey, Lawrence
MICHAEL F. GOODCHILD, University of California, Santa Barbara
STANLEY K. HONEY, The News Corporation, Ltd., Los Angeles
TERRENCE J. KEATING, Autometrics, Inc., Bangor, ME
MICHAEL D. MARVIN, MapInfo Corporation, Troy, NY
SARA L. McLAFFERTY, Hunter College, New York, NY
KAREN C. SIDERELIS, North Carolina Center for Geographic Information and Analysis, Raleigh
ROBERT TUFTS, TASC, Reston, VA
NANCY VON MEYER, Fairview Industries, Middleton, WI

Staff
THOMAS M. USSELMAN, Senior Staff Officer
JUDITH L. ESTEP, Administrative Assistant

Members Who Completed Their Terms on December 31, 1993
JOHN D. BOSSLER, The Ohio State University, Colubus, *Chairman*
ROBERT LEE CHARTRAND, Naples, Florida (retired, Congressional Research Service)
DONALD F. COOKE, Geographic Data Technology, Inc., Lyme, NH
GIULIO MAFFINI, SHL Systemhouse, Inc., Toronto, Canada
JOHN D. McLAUGHLIN, University of New Brunswick, Fredericton
BERNARD J. NIEMANN, JR., University of Wisconsin, Madison
GERARD RUSHTON, University of Iowa, Iowa City

BOARD ON EARTH SCIENCES AND RESOURCES

FREEMAN GILBERT, *Chair*, Scripps Institution of Oceanography, La Jolla, CA
GAIL M. ASHLEY, Rutgers University, Piscataway, NJ
THURE CERLING, University of Utah, Salt Lake City
MARK P. CLOOS, University of Texas at Austin
NEVILLE G.W. COOK, University of California, Berkeley
JOEL DARMSTADTER, Resources for the Future, Washington, DC
DONALD J. DePAOLO, University of California, Berkeley
MARCO EINAUDI, Stanford University, Stanford, CA
NORMAN H. FOSTER, Independent Petroleum Geologist, Denver, CO
CHARLES G. GROAT, Louisiana State University, Baton Rouge
DONALD C. HANEY, Kentucky Geological Survey, Lexington
ANDREW H. KNOLL, Harvard University, Cambridge, MA
PHILIP E. LaMOREAUX, P.E. LaMoreaux and Associates, Inc., Tuscaloosa, AL
SUSAN LANDON, Thomasson Partner Associates, Denver, CO
MARCIA K. McNUTT, Massachusetts Institute of Technology, Cambridge
J. BERNARD MINSTER, University of California, San Diego
JILL D. PASTERIS, Washington University, St. Louis, MO
EDWARD C. ROY, JR., Trinity University, San Antonio, TX

Staff
JONATHAN G. PRICE, Staff Director
THOMAS M. USSELMAN, Associate Staff Director
WILLIAM E. BENSON, Senior Program Officer
KEVIN CROWLEY, Program Officer
BRUCE B. HANSHAW, Program Officer
ANNE LINN, Program Officer
LALLY A. ANDERSON, Staff Assistant
CHARLENE ANDERSON, Administrative Assistant
JUDITH L. ESTEP, Administrative Assistant
SHELLEY MYERS, Project Assistant

COMMISSION ON GEOSCIENCES, ENVIRONMENT, AND RESOURCES

M. GORDON WOLMAN, The Johns Hopkins University, *Chairman,* Baltimore, MD
PATRICK R. ATKINS, Aluminum Company of America, Pittsburgh, PA
EDITH BROWN WEISS, Georgetown University Law Center, Washington, DC
JAMES P. BRUCE, Canadian Climate Program Board, Ontario
WILLIAM L. FISHER, The University of Texas at Austin
EDWARD A. FRIEMAN, Scripps Institution of Oceanography, La Jolla, CA
GEORGE M. HORNBERGER, University of Virginia, Charlottesville
W. BARCLAY KAMB, California Institute of Technology, Pasadena
PERRY L. McCARTY, Stanford University, Stanford, CA
JUDY L. MEYER, University of Georgia, Athens
RAYMOND A. PRICE, Queen's University at Kingston, Ontario, Canada
THOMAS C. SCHELLING, University of Maryland, College Park
ELLEN K. SILBERGELD, Environmental Defense Fund, Washington, DC
STEVEN M. STANLEY, The Johns Hopkins University, Baltimore, MD
VICTORIA J. TSCHINKEL, Landers and Parson, Tallahassee, FL
WARREN WASHINGTON, National Center for Atmospheric Research, Boulder, CO

Staff
STEPHEN RATTIEN, Executive Director
STEPHEN D. PARKER, Associate Executive Director
MORGAN GOPNIK, Assistant Executive Director
JEANETTE SPOON, Administrative Officer
SANDI FITZPATRICK, Administrative Associate
ROBIN ALLEN, Senior Project Assistant

The National Academy of Sciences is a private, nonprofit, self-perpetuating society of distinguished scholars engaged in scientific and engineering research, dedicated to the furtherance of science and technology and to their use for the general welfare. Upon the authority of the charter granted to it by the Congress in 1863, the Academy has a mandate that requires it to advise the federal government on scientific and technical matters. Dr. Bruce Alberts is president of the National Academy of Sciences.

The National Academy of Engineering was established in 1964, under the charter of the National Academy of Sciences, as a parallel organization of outstanding engineers. It is autonomous in its administration and in the selection of its members, sharing with the National Academy of Sciences the responsibility for advising the federal government. The National Academy of Engineering also sponsors engineering programs aimed at meeting national needs, encourages education and research, and recognizes the superior achievements of engineers. Dr. Robert M. White is president of the National Academy of Engineering.

The Institute of Medicine was established in 1970 by the National Academy of Sciences to secure the services of eminent members of appropriate professions in the examination of policy matters pertaining to the health of the public. The Institute acts under the responsibility given to the National Academy of Sciences by its congressional charter to be an adviser to the federal government and, upon its own initiative, to identify issues of medical care, research, and education. Dr. Kenneth I. Shine is president of the Institute of Medicine.

The National Research Council was organized by the National Academy of Sciences in 1916 to associate the broad community of science and technology with the Academy's purposes of furthering knowledge and of advising the federal government. Functioning in accordance with general policies determined by the Academy, the Council has become the principal operating agency of both the National Academy of Sciences and the National Academy of Engineering in providing services to the government, the public, and the scientific and engineering communities. The Council is administered jointly by both Academies and the Institute of Medicine. Dr. Bruce Alberts and Dr. Robert M. White are chairman and vice chairman, respectively, of the National Research Council.

PREFACE

The Mapping Science Committee serves as a focus for external advice to the federal agencies on scientific and technical matters related to spatial data handling and analysis. The purpose of the committee is to provide advice on the development of a robust national spatial data infrastructure for making informed decisions at all levels of government and throughout society in general.

Within the context of the above mission statement, the committee issued a report in the Spring of 1993, *Toward a Coordinated Spatial Data Infrastructure for the Nation*, which articulated its vision on how spatial information handling might best be approached from an organizational perspective. There are, of course, many specific issues that are raised when one examines what a National Spatial Data Infrastructure (NSDI) encompasses. The committee is undertaking a series of focused studies to examine individual components of the NSDI. This study on partnerships within the NSDI was verbally requested by the Federal Geographic Data Committee (operating under the aegis of the Office of Management and Budget) at a joint meeting with the Mapping Science Committee on February 2, 1993.

A study which addresses the criteria for determining priority geographic data is companion to this study on partnerships. Together, they set the stage for the NSDI which will take our nation into the twenty-first century.

The committee wishes to thank Lisa Warnecke for the compilation of state legislation and authorities concerning spatial data and of the coordination mechanisms within each state.

CONTENTS

1 EXECUTIVE SUMMARY ... 1
RECOMMENDATIONS ... 3
NOTE ... 4

2 THE NATIONAL SPATIAL DATA INFRASTRUCTURE ... 5
BACKGROUND ... 5
PREVIOUS REPORTS ... 8
Spatial Data Needs: The Future of the National Mapping Program, 8; *Toward a Coordinated Spatial Data Infrastructure for the Nation,* 8
DISCUSSION ... 9
NOTES ... 11

3 THE CONCEPTS OF FEDERAL/STATE PARTNERSHIPS ... 12
CREATION AND MAINTENANCE OF SPATIAL DATA SETS ... 12
SPATIAL DATA STEWARDSHIP PRINCIPLES ... 13
ECONOMICS OF THE NSDI ... 14
The Costs of NSDI, 15; *Minimizing Costs Through Partnerships,* 16
NOTES ... 18

4 THE PARTNERSHIP MODEL ... 19
KEY ELEMENTS OF PARTNERSHIPS ... 19
IMPEDIMENTS TO PARTNERSHIPS ... 22
Formulae for Cost Sharing, 22; *Compromised Standards,* 22; *Cost Recovery,* 23; *Federal Procurement,* 23; *Focused Coordination,* 24
ROLES IN PARTNERSHIPS ... 24
NOTES ... 26

5 RECOMMENDATIONS AND CONCLUSION ... 27
KEY ELEMENTS OF A PARTNERSHIP ... 27
RECOMMENDATIONS ... 27
CONCLUSION ... 28

APPENDICES

APPENDIX A: RELATED ACTIVITIES AND DEVELOPMENT OF PARTNERSHIPS 33
    NOTES 35

APPENDIX B: COOPERATION/PARTNERSHIP MODELS IN EXISTENCE TODAY 37
    BUREAU OF THE CENSUS—STATE DATA PROGRAM 37
    NATIONAL GEODETIC SURVEY 38
    SOUTH CAROLINA WATER RESOURCES
        COMMISSION 39
    MARYLAND DIGITAL ORTHOPHOTO PROGRAM . 41

APPENDIX C: EXAMPLE MEMORANDUM OF UNDERSTANDING 43

APPENDIX D: STATE GEOGRAPHIC INFORMATION AUTHORIZATIONS AND COORDINATION—SUMMARY 47
    STATE STATUTES 47
    EXECUTIVE ORDERS 48
    MEMORANDA OF UNDERSTANDING 48
    DISCUSSION 49
    STATE COMPILATION 52

ACRONYMS 113

Promoting the National Spatial Data Infrastructure Through Partnerships

1
EXECUTIVE SUMMARY

Cooperation and partnerships for spatial data activities among the federal government, state and local governments, and the private sector will be essential for the development of a robust National Spatial Data Infrastructure (NSDI).

The NSDI is the total ensemble of available geographic information that describes the arrangement and attributes of features and phenomena on the Earth, as well as the materials, technology, and people necessary to acquire, process, store, and distribute such information to meet a wide variety of needs. The twenty-first century will see geographic information transported from remote nodes using computer networks to support decision making throughout the nation. The National Information Infrastructure (NII) will provide the technology infrastructure to make this possible. There are vast amounts of spatial data ready to move across the information superhighways today. Timely use of these data would be difficult due to ill-defined format, quality, and accuracy. National or regional decision making would be severely impaired because most data sets are not adequately characterized. This is to be contrasted by the fact that the NII may well be the most important technology needed to facilitate a coordinated NSDI.

The Mapping Science Committee (MSC) has recommended[1] that the NSDI be developed to a level that would support the needs of the nation. The costs of creating and maintaining digital spatial data are high, so it is particularly important that spatial data collection not be duplicated, and that data be shared to fully realize its potential benefits. Largely for these reasons, the National Performance Review (prepared under the guidance of Vice President Gore) urged the formation of spatial data partnerships between federal agencies, state and local governments, and the private sector. After examining the pros and cons of several current spatial data programs that involve partnerships, the MSC agrees that a partnership model, the subject of this report, is an excellent approach for enhancing the NSDI. A companion report (in preparation—*The National Spatial Data Infrastructure: The Data Foundation*) suggests a rationale for identifying the principal elements needed to create the data component of the NSDI.

The focus of partnership arrangements within this report is on federal and state agencies. However, the MSC recognizes that a large volume of spatial data is created and used by local governments throughout the nation. Recently a number of state geographic information councils have been established to coordinate spatial data activities within the respective states. Such councils can also encourage partnerships between state and local government agencies, coordinate arrangements between state agencies and the private sector, and provide points of contact for partnerships with the federal government organizations. Although many states have geographic information councils, they are not universal. The MSC agrees with the recommendation of the Federal Geographic Data Committee (FGDC) in its strategic plan to help form or strengthen these state geographic information councils. The principles developed in this report should be transferrable to a wide variety of spatial data partnerships.

The committee identified several key elements during the course of this study that should be common to future partnerships. These include the following:

- **Shared Responsibilities.** The parties to a partnership should have a formal agreement that defines each party's responsibilities in the activity.
- **Shared Commitment.** The costs of the activity should be shared between the parties according to some agreed formula.
- **Shared Benefits.** Each party to the activity should derive some benefit that is consistent with its mandated role as an agency.
- **Shared Control.** Decision-making control of the partnership should be divided between the participants.

In addition, the committee believes that the following conditions should be emphasized in the formation of NSDI partnerships:

- Benefits of spatial data partnerships must be evaluated for the entire national community of spatial data users, not merely for the agencies participating in the partnership.

- The contribution of a spatial data partnership to the wider objectives of NSDI must be considered in its design and management.
- Data of known quality are an important factor in the value of any investment in spatial data. Potential users will be confident using data only if they know the data are reliable.
- Stewardship is a key concern in reaping the benefits of investment in any spatial data partnership; the organization closest to the source of the data should be best able to maintain the data.
- An essential element of the NSDI partnership model must be a commitment to support the partnership as part of an ongoing program. Long-term commitments will help ensure that data are maintained and that mutual trust in a partner's ability to meet respective needs will be achieved.

RECOMMENDATIONS

1. The size and diversity of the federal establishment suggest that viable partnerships will require focal points within the federal government for coordinating data production and partnership activities. The range of alternatives to consider should include regional coordination staff and coordinating positions within organizations responsible for spatial data production. Stewardship responsibilities for base (framework) data sets specific to geographic areas should be encouraged. The practicality of a data stewardship certification program should be studied. The clearinghouse function should catalogue data available through data stewards.

2. Clear guidelines for cost sharing in partnerships need to be developed. The formulation of such guidelines should be one component of the FGDC's role in NSDI. Guidelines should reflect the responsibility of the federal government to address and fund the nation's interest in the NSDI.

3. It is imperative that states and other organizations be involved in the standards development process and that only standards essential to NSDI objectives be required of partnership agreements. Promulgation and maintenance of standards is an important component of the FGDC's role in NSDI; standards must not be compromised in the formation of partnerships between state and federal agencies.

4. Incentives are needed to encourage partnerships that are designed to maximize use and benefits to the broader user community. Such incentives could be provided through the monitoring and coordinating roles of the FGDC and state geographic information councils.

5. The Federal Geographic Data Committee should investigate the extent to which federal procurement rules (and future revisions resulting from the National Performance Review) are an impediment to the formation of spatial data partnerships, and identify steps that can be taken to ease them.

The partnership model with data stewardship responsibilities at all levels of government represents a fundamental shift in the way the nation develops and supports the NSDI. Federal agencies will devote more resources to coordination and less to data production. States and other government entities will continue to expand their data production roles to support national needs. It is consistent with the recognition that today data are distributed across the nation at all levels of government and the public. The rapidly expanding communication network (the NII or information superhighway) will be the conduit to bring these data sets together. The FGDC, federal agencies, and state geographic information councils in concert will form a critical component of the infrastructure that ensures logical consistency and availability of framework and other spatial data to carry the nation into the next century.

NOTES

Toward a Coordinated Spatial Data Infrastructure for the Nation (1993). Mapping Science Committee, National Research Council, National Academy Press, Washington, D.C., 171 pp.

2
THE NATIONAL SPATIAL DATA INFRASTRUCTURE

BACKGROUND

It is generally accepted that in the early 1960s, the United States moved from being an industrial society to being an information society. This information society depends upon spatial (geographic) data and information. Until recently, maps (usually in paper form) have been a mainstay for a wide variety of applications and decision making.

This is changing as more spatially referenced data and information on a wider variety of topics or themes (e.g., population, land use, economic transactions, hydrology, agriculture, climate, soils) are being produced, stored, transferred, manipulated, and analyzed in digital form.

Several factors have contributed to the advancement of digital technology for collecting, handling, and processing spatial data. Perhaps the most important are the relative ease with which digital spatial data can be edited and updated (no more handwritten notes on paper maps); improved integration of operations between administrative departments within agencies (departments share and contribute to spatial data as a common resource); data management and storage; more effective analysis and decision making (manual analysis of paper maps is exceedingly tedious and costly); and faster access to current data (changes are available to all users in near real time).

With the current emphasis on digital spatial data, new products (representing the conversion of paper map information, the enhancement of that information, and the collection of new data) are appearing with greater frequency. With this increased production comes the potential for substantial duplication of effort, as virtually identical digital products appear from different agencies to satisfy their often very specific needs. The costs of creating and maintaining digital spatial data are high, so it is particularly important that data created at considerable cost and effort be shareable, that costly data collection not be duplicated, and that the collected data be fully utilized to realize all of their potential benefits.

The next decade will see rapid and large-scale investment in communications technology as the nation moves to exploit the full potential of the information age. Recent actions by the federal government, including the passage of the act calling for a National Information Infrastructure (NII), as well as announced plans for private and public investment, make it clear that within a few years an unprecedented capability will exist for sharing of data along "electronic superhighways." Investment in digital communications technology has been likened to the national investment in the interstate highway network in the 1950s and 1960s, which spawned a major restructuring of U.S. society. Already some 10 million users of the Internet research network communicate nationally and internationally at megabit speeds. Many are predicting that within the next 10 years we will see a similar development in the consumer marketplace.

The high-speed communication networks will be essential for widespread access and sharing of spatial data. Spatial data tend to be voluminous, and sharing has traditionally been difficult at the communication speeds and bandwidths that were available in the past. Standards were often absent, or confusing at best, and it was often more cost-effective to communicate by mailing a paper map and redigitizing it rather than confront the problems of digital format conversion. In the last few years the research community has begun to develop effective methods for describing data quality and other aspects of data that potential users must know if they are to be able to assess the potential value of someone else's spatial data for their own use.

A major challenge over the next decade will be to increase the use of spatially referenced data to support a wide variety of decisions at all levels of society. Using an effective, efficient, and widely accessible NII, spatial data could be readily transported and easily integrated both thematically (e.g., across environmental, economic, and institutional data bases) and hierarchically (e.g., from local to national and eventually to global levels). Transparent access to myriad data bases could provide the information for countless applications, e.g., facility management, real estate transactions, taxation, land-use planning, transportation, emergency services, environmental assessment and monitoring, and research. Work on these applications could take place in schools, offices, and homes across the nation. In addition, these activities will lead to new value-added services and market opportunities in emerging spatial information industries.

In consideration of these challenges and opportunities, the Mapping Science Committee (MSC)[2] conceptualized a National Spatial Data Infrastructure (NSDI) as "the total ensemble of geographic information at our disposal that describes the arrangement and attributes of features and phenomena on the Earth, as well as the materials, technology and people necessary to acquire, process, store and distribute such information to meet a wide variety of needs." In its broadest sense, the infrastructure also includes the cultural, environmental, economic, political, legal, and educational values and institutions that support, facilitate, and shape its character, including the forms in which spatial data are represented and utilized throughout society. The MSC's concept of the NSDI is illustrated in Figure 1.

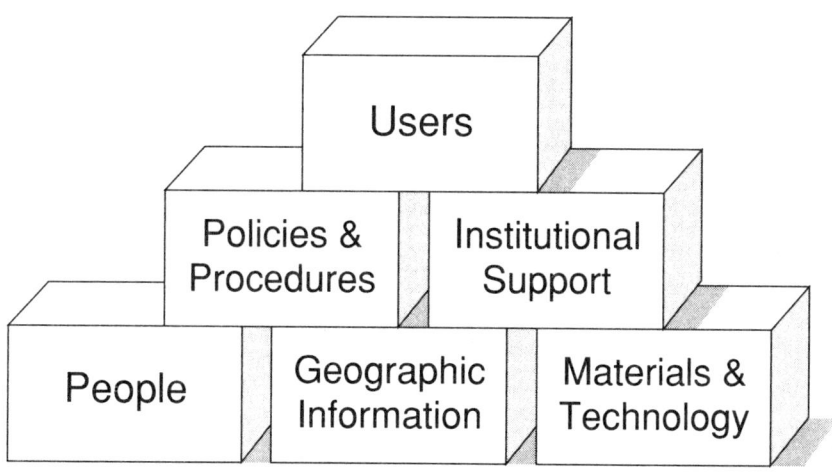

Figure 1. NSDI building blocks.

PREVIOUS REPORTS

Spatial Data Needs: The Future of the National Mapping Program

In 1990, the MSC issued a report entitled *Spatial Data Needs: The Future of the National Mapping Program*.[3] The report resulted from numerous written reports and briefings provided by the U.S. Geological Survey (USGS). The vision of a cartographic enterprise and its sequel, the NSDI was developed, and recommendations pertaining to the National Mapping Division (NMD) of the USGS were provided. One of the most important recommendations made in the report is that the future role of the NMD should focus on coordination of certain NSDI functions in addition to production of spatial data. With the establishment of the Federal Geographic Data Committee (FGDC) by the Office of Management and Budget (OMB) to address coordination and standards objectives, NMD will be able to focus on supporting the operational aspects of NSDI development. In keeping with the recommendations, this current report addresses one aspect of the NSDI—the concept of federal/state partnerships to achieve production objectives.

Toward a Coordinated Spatial Data Infrastructure for the Nation

After assimilating information from 12 federal agencies, the MSC produced a report entitled *Toward a Coordinated Spatial Data Infrastructure for the Nation*.[2] This report defined the elements of the NSDI and provided a conceptual framework for an enhanced spatial data infrastructure for the nation. The recommendations of that report follow:

1. Effective national policies, strategies and organizational structures need to be established at the federal level for the integration of national spatial data collection, use, and distribution.

2. The Federal Geographic Data Committee (FGDC), which operates under the aegis of the Office of Management and Budget (OMB), should continue to be the working body of the agencies to coordinate the interagency program as defined in

OMB Circular A-16. However, strengthening the charter is needed to:

- expand the development and speed the creation and implementation of standards (content, quality, performance and exchange), procedures, and specifications for spatially-referenced digital data; and
- create a series of incentives, particularly among federal agencies, that would maximize the sharing of spatial data and minimize the redundancy of spatial data collection.

3. Procedures should be established to foster ready access to information describing spatial data available within government and the private sector through existing networks, thereby providing on-line access by the public in the form of directories and catalogs.

4. A spatial data sharing program should be established to enrich national spatial data coverage, minimize redundant data collection at all levels, and create new opportunities for the use of spatial data throughout the nation. Specific funding and responsibilities of federal agencies should be identified by OMB, and the FGDC should coordinate the across-agency aspects of the program.

In keeping with this last recommendation, this present report expands the ideas of cooperation and partnerships between federal and state spatial data activities within the NSDI.

DISCUSSION

The NSDI is in existence today. It continues to evolve as the FGDC continues to define its role and responsibility for the nation's geographic data of the future. The Departments of Interior, Agriculture, Commerce, and Defense have been traditionally the major suppliers of spatial data. Most other departments and independent agencies are users, and sometimes suppliers, of spatial data. At the state level, over two-thirds of the states have active geographic information councils, geographic information systems (GIS) coordination groups, or similar organizations responsible for coordinating spatial data; data collection activities are dispersed among the various agencies within a state. At the local level, most counties, municipalities, and utility companies are pursuing the collection and use of spatial data to address their information needs for management and planning.

There seemed to be a lack of coherent direction for the NSDI in the past, and the infrastructure often appeared chaotic. At all levels of spatial data handling, organizations are confronted by myriad problems, confusing policies, and even disincentives to coordinate their activities. Many of these problems are addressed in the previous reports of the MSC. The FGDC has made significant advancements in the past two years by effective communications of the NSDI challenge through newsletters, magazines, and professional journals. The FGDC has hosted conference forums and opened communications with state geographic information councils. This activity has served to heighten the awareness of NSDI and the need for broad-scale coordination to meet NSDI objectives. The metadata and spatial data transfer standards developed over the past few years are examples of the standards framework needed to enhance the NSDI. Without them, the ability to bring diverse data sets together would be severely impaired.

The federal government agencies and states will play an important part in building the data sets that go into the NSDI. Through standards development, the data structures have been minimally defined at least. The data framework needed to support NSDI will be addressed in a companion report *The National Spatial Data Infrastructure: The Foundation Data*, which is in preparation. This report promotes a concept that partnerships are valid means to develop these data sets and are essential to achieve longer-term NSDI objectives.

A comprehensive partnership program no doubt will have its inherent problems, but it will provide the opportunity to build the NSDI rapidly. With this in mind, the body of this report develops the partnership concepts. It begins with a discussion of the costs associated with the NSDI and how these costs could be minimized through partnerships. Then, the critical success factors for a federal/state partnership model for spatial data are presented; the same factors would apply to partnership models between any organizations. Partnership initiatives are being promoted currently as a valid approach to develop the information infrastructure. A review of relevant literature (Appendix A) clearly reveals this trend as well as the inherent challenges to make it all work. The factors were derived from several data sharing programs currently in existence (Appendix B). Impediments to this model and roles of the various organizations in such partnerships are also discussed. The report concludes with a number of recommendations to further develop these spatial data partnerships.

NOTES

[1] *Realizing the Information Future: The Internet and Beyond*, 1994, Computer Science and Telecommunications Board, National Research Council, National Academy Press, Washington, D.C., 320 pp.

[2] *Toward a Coordinated Spatial Data Infrastructure for the Nation*, 1993, Mapping Science Committee, National Research Council, National Academy Press, Washington, D.C., 171 pp.

[3] *Spatial Data Needs: The Future of the National Mapping Program*, 1990, Mapping Science Committee, National Research Council, National Academy Press, Washington, D.C., 78 pp.

3
THE CONCEPTS OF FEDERAL/STATE PARTNERSHIPS

THE CREATION AND MAINTENANCE OF SPATIAL DATA SETS

Historically, the federal spatial data infrastructure has been managed as a set of discrete mapping responsibilities within several federal agencies. The data management role of federal agencies has been that of data stewards for large homogeneous data sets (more often in paper map form rather than digital data sets). Examples of such data sets include the USGS 7.5 minute quadrangle maps, Soil Conservation Service soils maps, census geography maps from the U.S. Bureau of the Census, and the Fish and Wildlife Service national wetland inventory maps. As part of their mandates, federal agencies have collected and published data on maps that were then distributed to all levels of government and the private sector. Relationships with states have been largely limited to ad hoc or cooperative projects, with the states generally taking on the role of suppliers of data to the federal agencies.

This set of historical arrangements has many problems, both obvious and subtle. The activities of the federal government have been established largely by legislative mandate, and although these may be initially tied to need, it has been difficult to keep up with changing technologies and changing requirements in the user community. Mandates sometimes lead to redundancy, as legislators require different agencies to collect and maintain similar or even identical data. Costs are difficult to identify, quantify, or control, and the government finds it increasingly difficult to maintain established programs in the face of constantly increasing pressures on the federal budget. Finally, programs of spatial data creation have often been slow and inadequately funded for data maintenance. As a result, data are often so out of date that their value is seriously compromised. These problems are increasingly evident when the state data programs are viewed in the same context and as part of the NSDI.

The role of state and local governments within the overall context of the NSDI is changing. The federal coordination of surveying, mapping, and related spatial data activities are defined in OMB Circular A-16. The most recent version of Circular A-16 (October 1990) added a major objective of developing a national digital spatial information resource with the involvement of federal, state, and local governments and the private sector. The mechanism for the involvement of state and local governments and the private sector has yet to be established. This problem was specifically recognized in Executive Order 12906 (signed by President Clinton on April 11, 1994):

> The Secretary [of the Department of the Interior], under the auspices of the FGDC, and within 9 months of the date of this order, shall develop, to the extent permitted by law, strategies for maximizing cooperative participatory efforts with State, local, and tribal governments, the private sector, and other non-federal organizations to share costs and improve efficiencies of acquiring geospatial data consistent with this order.

SPATIAL DATA STEWARDSHIP PRINCIPLES

Organizations that build and maintain spatial data have a vested interest in the quality of their data when the success of critical missions depends on the accuracy and availability of the data. This suggests that data stewardship roles may be served best by organizations that collect data for the purpose of meeting specific operational missions. These spatial data stewards have commitments to their own organization as well as obligations to meet the needs of partners throughout all levels of government. The mission of spatial data stewards could be expanded to meet data needs of multiple organizations as well as their own business needs. Currently, a major mission of the USGS is one of data collection. The stewardship concept introduced here is different from the data steward role of the USGS in that business needs other than a data collection mission also drive the development and maintenance of data.

The challenge faced by all levels of government is to place data stewardship responsibility as close to the data originator as possible while maintaining an effective national infrastructure. Partnership agreements may result in data stewardship responsibilities shifting between federal, state, and local agencies. Federal agencies should take on new coordination roles to ensure that data are available to meet a growing national need. State and local government agencies with mandates and other needs for certain data should be responsible for building and maintaining these data. This is occurring on an informal basis with many states now but will take on entirely new significance in the future as more organizations depend on others for data.

One clear benefit of moving data responsibility closer to the source and in line with the business needs of organizations is that maintenance of the data is usually an integral part of their day-to-day activities. As an example, the Bureau of Land Management (BLM) has the federal responsibility for the Public Land Survey System (PLSS). But in many states, after years of requesting the necessary funds, BLM is still unable to meet its objectives for modernization of its Geographic Coordinate Data Base. However in Washington State, for example, the State Land Surveyor has the responsibility to maintain current records of surveys and to make them available to land surveyors on demand. The State Land Surveyor also has the responsibility to manage and make PLSS data available to GIS customers throughout the state. As new surveys are recorded, the PLSS section corners are added to the data base. A partnership opportunity clearly exists that would establish data stewardship responsibility close to the source of the data and provide opportunities to meet both state and federal objectives.

ECONOMICS OF THE NSDI

One of the strongest arguments for partnerships is their critical role in reducing the overall costs of the enterprise. After 30 years of experience in spatial data handling, there can be no remaining doubt about one fundamental truth—it is expensive. Spatial data bases are expensive to create and often more expensive to maintain.

Why is spatial data handling so expensive, and where does the money go? How can partnerships help to reduce costs, and what is needed to ensure that economies are realized? This section looks briefly at these issues and at the ways in which federal/state partnerships in particular might help to minimize the costs of developing the NSDI.

The Costs of NSDI

Spatial data require specialized data collection systems, specialized hardware and software, and specialized training and education of the necessary support staff. All of these factors contribute to the overall cost of spatial data and the NSDI. Although GIS software development is a relatively small part of the overall electronic data processing industry, and few software companies claim a GIS business of more than $100 million annually, it is estimated that worldwide total annual expenditures on *all* aspects of spatial data handling are on the order of $10 billion.[1] This figure should include every aspect of data base creation and use, from investments in mapping satellites, geodesy and positioning systems, through airborne photography, digitizing, data base design, software and hardware acquisition, and analysis, to specialized training. The cost of data input by digitizing and scanning (data conversion in the terminology of activities related to automated mapping and facilities management) is estimated by the Environmental Systems Research Institute to be $4.5 billion[2] annually. The OMB recently surveyed federal spending on geographic data activities and found it to be about $4.4 billion (FY1994).[3]

These figures reflect current practice. As such, they miss much of the potential future role of spatial data in society, and the demands that are driving the interest in geographic information. We have only just begun to tap the potential that is illustrated by the range of consumer products based on digital spatial data bases that are now entering the market such as consumer GPS, visitor street guides to cities, kiosk vending of custom maps, and applications of IVHS.

As with any new field, there is no doubt inefficiency and waste in the current practices of spatial data handling. There is duplication of effort when the same map is digitized or scanned more than once, adding to the already high cost of converting spatial data to digital form. Duplication is hard to document, as it is rarely in anyone's interest to report when it occurs, and it would be impossible to come up with reliable estimates of the costs of duplication on an annual, nationwide basis. It occurs because there is little knowledge of what data sets exist or because the data might not be at appropriate detail or accuracy to be shared. The need to establish a clearinghouse has been recommended by the MSC[4] and is being developed as a prototype by the FGDC. Ironically, it can cost money to avoid duplication by developing data indexes, improving communication through conference attendance, and building data sets to meet other organizational requirements.

As the field matures, some problems will disappear as communication becomes easier and the level of general information on activities improves. But that will leave the more difficult problems. Duplication is obvious when two agencies each digitize the same map. It is much more difficult to persuade two agencies to collaborate in the collection of raw data, or in analysis, especially when collaboration requires some loss of autonomy or disruption of traditional professional boundaries. It is easy, also, to overestimate the occurrence of duplication by overlooking the subtler aspects of spatial data needs. Two agencies may both need wetlands data, but for different purposes, using different definitions and different levels of precision and accuracy. In such cases, reduction of duplication can be a lengthy and difficult business. Yet the rewards can be enormous. Although digitizing is expensive, and it is important to avoid obvious cases of duplication, the potential economies that can be achieved by collaboration and coordination in the total mapping effort are equally significant.

One of the major barriers to realizing potential economies from decreased duplication is the lack of agreed-upon standards, particularly data content standards, as previously discussed. The problems with different software systems should be allayed with adherence to the Spatial Data Transfer Standard (SDTS), which is now a Federal Information Processing Standard (FIPS-173). Specific content, accuracy, and metadata standards present much of the challenge facing the spatial data community. The various subcommittees of the FGDC are currently addressing many of these standard issues from the federal perspective. However, the state and local governments and the private sector need to be active participants in the standards development process.

Minimizing Costs Through Partnerships

It may appear that partnerships are exactly the wrong way to reduce costs in an expensive enterprise like the NSDI. It costs money to organize and facilitate partnerships. Partnerships can seem cumbersome when compared to a lean, efficient organization that carries out its mandate as inexpensively as possible. However, no matter how internally efficient an organization is, it is inefficient in the broader perspective if it duplicates the products of others. In practice, we believe that partnerships can reduce the long-term cost of NSDI in three major ways.

First, partnerships are an effective way of achieving consensus. Instead of each agency acting independently, partnerships create a sense of shared responsibility for the product and its use. Partnerships broaden the basis of support for projects and help to ensure that they survive to meet the needs of society. Partnerships between federal, state, regional, and local governments act to dispel the perception that one level of government ignores the others or knows better. In economic terms, partnerships broaden the resource base by sharing costs while enhancing the benefits of spatial data.

Second, partnerships can encourage a clear division of responsibilities even when the data needs are shared. Historically, responsibility for spatial data has been divided between different levels of government in the United States on the basis of map scale and types of data. For example, the federal government has concentrated on producing maps at 1:24,000 scale and smaller while leaving larger-scale mapping to local and state governments and the private sector. States have mapped themes that match their areas of responsibility, such as transportation. *These familiar divisions are becoming confused because of the radical changes resulting from the introduction of digital technology.* For example, the federal government is supporting a digital orthophoto quarter-quad (DOQ) program with spatial resolution and positional accuracy higher than that of the 1:24,000 scale mapping programs. GIS is being used by all levels of government to take advantage of these new data types and to integrate data from a wide variety of sources. In a world of high-speed communications and distributed data bases, we may need entirely new concepts of ownership of data, or responsibility for its creation. This may take the form of a division of responsibility along entirely new lines, with the federal government responsible for data standards and quality control, and state and local governments responsible for data collection and maintenance.

Third, division of responsibilities within partnerships can promote investment so that we develop entirely new ways of reducing costs. Salaries account for by far the largest share of the costs of spatial data, whether they are paid to digitizer operators, programmer analysts, or field workers. The most effective ways of reducing those costs lie in better technology and better training. Spatial data handling and GIS have grown to the point where creative strategies are needed to promote new methods and better technologies and better development of human resources; however, no agency or level of government has assumed a leadership role in such developments. Partnerships could foster a sense of shared responsibility between all levels of government, the educational sector, and private industry. We need partnerships that foster more efficient data collection activities while at the same time fostering a more productive and responsive human resource sector.

NOTES

[1] The $10 billion figure was arrived at in the 1993 MSC report *Toward a Coordinated Spatial Data Infrastructure for the Nation*.

[2] This estimate was given by J. Dangermond (president, Environmental Systems Research Institute, Redlands, California) in a speech (sponsored by Texaco) in June 1992 in Houston, Texas.

[3] This financial information was collected in the summer of 1993 for those federal programs involved in acquisition, management, and dissemination of geographically referenced data. Only agencies with a minimum of $500,000 in relevant spending were required to submit information in response to OMB Bulletin 93-14.

[4] *Toward a Coordinated Spatial Data Infrastructure for the Nation*, Mapping Science Committee, National Research Council, National Academy Press, Washington, D.C., 171 pp. The clearinghouse was also explicitly mentioned in Executive Order 12906 (April 11, 1994).

4
THE PARTNERSHIP MODEL

KEY ELEMENTS OF PARTNERSHIPS

Previous sections of this report have identified some of the advantages of partnerships and the potential roles they could play in development of the NSDI. In this section, attention is directed toward the identification of a model of partnerships and their general characteristics. Issues arising from partnerships are discussed, as well as potential impediments to their success.

A partnership model is envisioned that promotes long-term organizational commitments to build, maintain and manage data for a robust NSDI. The model could replace many of the mechanisms currently in place for developing and managing large spatial data sets and would facilitate long-term maintenance and availability of valuable spatial data. The model may well be the only approach that will attract the participation of states at the levels necessary to ensure the long-term viability of the NSDI.

For the purposes of this discussion, a partnership is defined as *a joint activity of federal and state agencies, involving one or more agencies as joint principals focusing on geographic information*. Local agencies and the private sector may also be involved. The key elements of a partnership follow.

- **Shared Responsibilities**. The parties to a partnership will have made a formal agreement that defines each party's responsibilities in the activity, in the form of a memorandum of understanding, contract, or other binding document.
- **Shared Costs**. The costs of the activity will be shared between the parties according to some agreed formula.

- **Shared Benefits**. Each party to the activity will derive some benefit that is consistent with its mandated role as an agency. In addition, benefits from the partnership will likely accrue to parties outside the partnership and to society at large. This is especially important when the partnership is seen as contributing to the evolution of the NSDI.
- **Shared Control**. Decision-making control of the project will be divided between the participants.

Each of these elements is important whether the partnership is merely an agreement between two agencies to carry out an activity to their mutual benefit, or whether it is fostered and encouraged as a contribution to NSDI. However, in the latter case certain other conditions are important.

- **Benefits.** A key aspect of the philosophy behind the idea of a NSDI is that spatial data are a national resource. If spatial data are created to satisfy the needs of one agency alone, or two or more agencies acting in partnership, then benefits to society at large may not be realized, and the result may be duplication of effort as each agency creates the spatial data that meet its own requirements. *Benefits of spatial data partnerships must be evaluated for the entire national community of spatial data users, not merely for the agencies participating in the partnership.* The secondary uses of data are becoming just as important as the primary purpose for which they were collected. This principle should form an important criterion in government's assessment and evaluation of any proposed partnership.
- **Design.** The design of a spatial data partnership must address the needs of potential users beyond the partner agencies. How will such users gain access to the data; how will the data be documented and catalogued so that they are easy to find; how can the concerns of secondary users be represented in the process of data base design; how can society benefit to the greatest extent possible from this investment in spatial data? *The contribution of a spatial data partnership to the wider objectives of NSDI must be considered in its design and management.*
- **Data Quality.** *Data quality is an important factor in the value of any investment in spatial data. Potential users will be confident using data only if they know the data are reliable*—if they are accurate, up to date, and consistent with their own documentation and metadata. Data that fail to

meet these criteria rapidly lose all the value that may have been invested in them.
- **Stewardship.** A previous section of this report argued that quality is best assured when data are maintained as close to the source as possible—that the agency responsible for collecting and building the data base is also the agency best able to maintain and assure its quality. In a federal/state partnership, it is likely that the state agency or agencies will be closer to the source. *Stewardship is a key concern in reaping the benefits of investment in any spatial data partnership; the agency closest to the source of the data is likely the agency best able to maintain the data.*

Ideally, there should be one data steward for any standard data set in any geographic area. The federal government cannot realistically police the program. It must, however, know where the data exist and be able to quickly assemble a variety of data to satisfy program objectives. It would be inconceivable that the federal government would enter into multiple partnership agreements for the same data set in a geographic area. A stewardship certification program along with directed funding and coordination with regional and local experts should be a vital component of a partnership program.
- **Sustained Relationship.** *An essential element of the NSDI partnership model must be a commitment to support the partnership as part of an ongoing program. Long-term commitments will help ensure that data are maintained and that mutual trust in a partner's ability to meet respective needs will be achieved.* Without the long-term partnership commitment, the NSDI could conceivably suffer from an endless string of hurry-up cooperative projects. The end result would be unnecessarily high administrative costs and a data base that is no more current than the present spatial data infrastructure.

Through the course of this investigation, many partnerships between organizations were reviewed. The above elements are considered to be essential to successful partnerships that pass the test of time. Primary consideration needs to be given to the agreements that formalize these relationships. Appendix C is an example of such an agreement. It contains all of the elements discussed above yet is simple and to the point. Expectations and roles of individual agencies are well defined. The agreement could serve as a basis for formulating future agreements between other organizations.

IMPEDIMENTS TO PARTNERSHIPS

All of the successful cooperation/partnerships models reviewed during this study (Appendix B) contain some partnership elements discussed above, and all have resulted in benefits that extend well beyond the concerns of the sponsoring agencies. At the same time, there are several significant impediments to the formation and success of partnerships. Some are unavoidable, but in other cases there are actions that could be taken that would reduce or even remove these impediments.

Formulae for Cost Sharing

One rational way of determining the contribution of each agency to a partnership activity's costs would be on the basis of benefits. A state agency would pay according to the benefits it derived; a federal agency should contribute according to the benefits that it, other federal agencies, and the nation as a whole derived from the activity. However, it is difficult to assess the benefits of spatial data in any but the narrowest range of applications. The benefits of a contribution to NSDI would be especially difficult to assess. In practice, the costs of a partnership are often divided equally, but a wide range of models have been used. *The lack of clear guidelines for cost sharing in partnerships is an impediment to their formation and success. The formulation of such guidelines should be one component of the FGDC's role in NSDI. Guidelines should reflect the responsibility of the federal government to address and fund the nation's interest in NSDI.* Without clear guidelines, it is difficult to avoid inconsistency.

Compromised Standards

If society is to reap the full benefits from an investment in spatial data, it is important that the data be created according to standards for both content and format. When partnerships are negotiated between levels of government, each party in the negotiation may have their own requirements on format, accuracy, and other technical aspects of the data. The result is often a compromise; the standards promulgated by the federal agency are broadened to meet the needs of the state. A nationwide coverage created by a series of partnerships with states can become so compromised that its eventual benefits are seriously eroded, and the result is a patchwork of different formats and accuracies. Although there are successful examples (e.g., the State Plane Coordinate system maintains a reasonable level of consistency across the nation), it is important that any program of partnerships include sufficiently strong incentives to maintain standards. *It is imperative that states be involved in the standards development process and that only those standards essential to NSDI objectives be required of partnership agreements. Promulgation and maintenance of these standards is an important component of the FGDC's role in NSDI; standards must not be compromised in the formation of partnerships.*

Cost Recovery

Some countries, notably the United Kingdom, have moved in recent years toward full recovery of the costs of collecting and maintaining spatial data. The United States, on the other hand, remains firmly committed to the notion that spatial data should be distributed at the cost of reproduction, at least at the federal level. Spatial data within the federal government are a public good, and are treated as a national resource and made available to all users at the least possible cost. One of the strongest arguments for this policy is its positive impact on the development and strength of industrial, commercial, and service sectors of U.S. spatial data activities. However, it creates little incentive for agencies creating spatial data to evaluate the broader need for the data, or to reduce cost through sharing, despite the importance of sharing as an underlying principle of NSDI. In the private sector, and in countries like the United Kingdom, such incentives are provided through the market mechanism. *Lack of incentive to evaluate the potential user base for spatial data, and to tailor data to maximize use and benefits in the broader community, is an impediment to partnerships and the evolution of NSDI. Such incentives could be provided through the monitoring and coordinating roles of the FGDC and state geographic information councils.*

Federal Procurement

The complexity and length of the federal procurement process[1] is a major impediment to the formation of creative partnerships with other levels of government. The federal government imposes rules that can make it very difficult to form contracts with other levels of government or to transfer money to them. *The FGDC should investigate the extent to which procurement rules (both federal and agency specific) are an impediment to the formation of spatial data partnerships, and identify steps that can be taken to ease them.*

Focused Coordination

State government must work with many federal agencies in order to initiate spatial data collection programs. Likewise, federal agencies desire to have focal points in state government for the purpose of implementing national programs. For example, the establishment of State Mapping Advisory Committees was in large part a desire to have a mechanism for states to channel their requirements to the USGS's NMD for the national mapping program. The state geographic information councils are the present day response by states to provide focal points for coordination of spatial data needs. *The size and diversity of the federal system suggests that for viable partnerships some action must be taken to provide focal points within the federal government for coordinating data production and partnership activities. The range of alternatives to consider should include regional coordination staff and coordinating positions within organizations responsible for spatial data production.*

ROLES IN PARTNERSHIPS

This section addresses the roles played by key agencies in the development of the NSDI—the FGDC, the various state coordinating groups and bodies for spatial data, and the NMD of the USGS.

• **Federal Geographic Data Committee**. The FGDC should have as a major goal that the nation derives the maximum possible benefit from partnerships. To do so, FGDC should facilitate and encourage the formation of partnerships; identify areas where partnerships might make a particularly beneficial contribution to NSDI; devise policies and guidelines to address the requirements of partnerships and the impediments that discourage them; ensure adequate concern for the needs of the wider user base; monitor and evaluate the success of partnerships; promulgate standards and encourage adherence; and formulate federal policy relating to NSDI. Planning for many of these components are specified in Executive

Order 12906.

The FGDC should manage the spatial data indexing or clearinghouse function and promote the data stewardship program. The FGDC should also have a major role in coordinating the development of partnership agreements. Because of geographic diversity, this role might be administered through regional FGDC activities.

- **State Geographic Information Councils.** Although in some cases less formally organized than the FGDC, the state councils are the logical NSDI focus within each state. As such, they should interact with the FGDC and echo its national concern for the NSDI at the state level. The state geographic information infrastructures and their laws and directives could affect their ability to respond to partnership initiatives. A discussion and synopsis of the results of a comprehensive listing of 100 state directives, including statutes, executive orders (by governors), and memoranda of understanding (MOUs) that mention geographic information and may directly influence the NSDI, is part of Appendix D. Many states have some form of authorization that provide opportunities for federal/state partnership development.

Statewide geographic information coordination groups exist in each of the 50 states. Appendix D provides a compilation, description and identification of the groups. Most of the groups are multiagency in focus and membership. Overall, there is a growing trend focusing on all geographic information and related technologies.

A generalized organizational diagram describing statewide geographic information organizations is presented in Figure 2. A council composed of stakeholders who are able to set policy and provide focused coordination is an essential component of the model. A geographic information association allows for broad membership participation and input to the council. A number of subcommittees representing special interests or charged with specific activities interact with the council and association. These are all important components to ensure that coordinated development of a spatial data infrastructure will occur. An effective organizational structure will also contribute to a long term stable partnership relationship.

- **National Mapping Division, USGS.** NMD is the logical federal agency to provide the lead technical support for NSDI for such activities as devising standards for data and metadata; researching methods for data sharing, enhancement of the benefits of spatial data and reductions in their cost; building networks, clearinghouses, catalogs, and other improved access mechanisms for NSDI; and researching and formulating improved models for the evaluation of benefits and sharing of costs. These NSDI missions are similar to those discussed in the committee's 1990 report.[2]

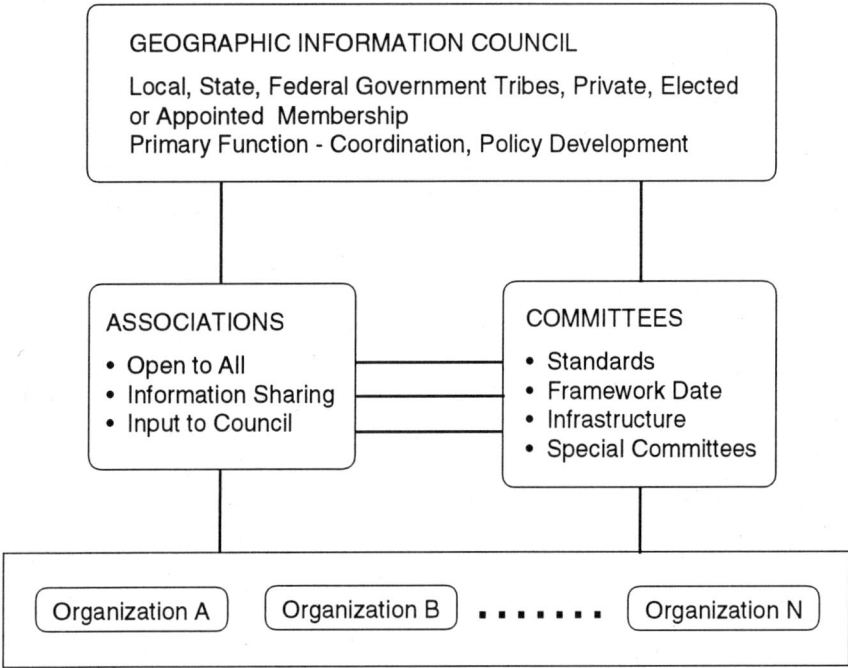

Figure 2. Generalized organizational diagram describing statewide geographic information organizations in 1993.

NOTES

[1]Many of the issues on federal procurement are addressed in *Creating a Government That Works Better & Costs Less: Report of the National Performance Review*, 1993, Vice President Al Gore, U.S. Government Printing Office, Washington, D.C., 168 pp.

[2]*Spatial Data Needs:The Future of the National Mapping Program*, 1990, Mapping Science Committee, National Research Council, 78 pp.

5
RECOMMENDATIONS AND CONCLUSION

KEY ELEMENTS OF A PARTNERSHIP

- Benefits of spatial data partnerships must be evaluated for the entire national community of spatial data users, not merely for the agencies participating in the partnership.
- The contribution of spatial data partnerships to the wider objectives of NSDI must be considered in its design and management.
- Data quality is an important factor in the value of any investment in spatial data. Potential users will be confident using data only if they know the data are reliable.
- Stewardship is a key concern in reaping the benefits of investment in any spatial data partnership; the agency closest to the source of the data is likely the organization best able to maintain the data.
- An essential element of the NSDI partnership model must be a commitment to support the partnerships as part of an ongoing program. Long-term commitments will help insure data are maintained and that mutual trust in the ability of the partners to meet respective needs will be achieved.

RECOMMENDATIONS

1. The size and diversity of the federal establishement suggest that viable partnerships will require focal points within the federal govern-

ment for coordinating data production and partnership activities. The range of alternatives to consider should include regional coordination staff and coordinating positions within organizations responsible for spatial data production. Stewardship responsibilities for base (framework) data sets specific to geographic areas should be encouraged. The practicality of a data stewardship certification program should be studied. The clearinghouse function should catalogue base data and their availability through data stewards.

2. Clear guidelines for cost sharing in partnerships need to be developed. The formulation of such guidelines should be one component of the FGDC's role in NSDI. Guidelines should reflect the responsibility of the federal government to address and fund the nation's interest in NSDI.

3. It is imperative that states and other organizations be involved in the standards development process and that only standards essential to NSDI objectives be required of partnership agreements. Promulgation and maintenance of standards is an important component of the FGDC's role in NSDI; standards must not be compromised in the formation of partnerships.

4. Incentives are needed to encourage partnerships that are designed to maximize use and benefits to the broader user community. Such incentives could be provided through the monitoring and coordinating roles of the FGDC and state geographic information councils.

5. The Federal Geographic Data Committee should investigate the extent to which federal procurement rules (and future revisions resulting from the National Performance Review) are an impediment to the formation of spatial data partnerships, and identify steps that can be taken to ease them.

CONCLUSION

In conclusion, the MSC finds that the FGDC, federal agencies, and state geographic information councils are making positive contributions to the evolution of the NSDI. The MSC further identifies the partnership

RECOMMENDATIONS AND CONCLUSION

model and approach as one of the key components of the enhanced NSDI. This report encourages the acceleration of partnership activities and identifies probable areas that will need attention to ensure success. The single largest challenge will be for many organizations to move from predominately production activities to accomplishing objectives through partnerships. This will necessarily mean that organizational tactical plans be reviewed and modified to achieve the objectives of a more robust NSDI.

APPENDIXES

APPENDIX A
RELATED INITIATIVES AND DEVELOPMENT OF PARTNERSHIPS

The promotion of the NSDI through partnerships represents but one dimension of a broader national thrust toward enhanced federal/state cooperation. Through legislation, governmental pronouncements, conference emphases, and media commentaries alike, the search for new and better ways of shaping a more efficient and responsive system for acquiring, maintaining, and distributing spatial data continues. Early in the new Administration, President Clinton and Vice President Gore stated, "Effective management of technology policy . . . requires an effective partnership between federal and state governments. The states have pioneered many valuable programs to accelerate technology development and commercialization."[1]

Several organizations and reports have recently focused on the high potential of such collaborative efforts.

- The National Association of State Information Resource Executives[2] has interacted directly with the OMB in a positive effort to establish effective partnerships to change paradigms and prototype new approaches. Case studies with proposed solutions were identified—from such states as Florida, Kentucky, Mississippi, Ohio, Oregon, South Carolina, and Texas—across a spectrum of applications.
- The stated goal of the report prepared by the State Information Policy Consortium[3] was to provide "a scenario in which technology and information can be used to re-engineer and streamline government operations at all levels." The consortium consists of representatives from the Council of State Governments, the National Conference of State Legislatures, and the National Governors' Association.

- A study issued by the National Academy of Public Administration[4] further reflects the belief that "information is pivotal to the vitality and productivity of government services and the nation's competitiveness."
- In Vice President Gore's report[5] on the National Performance Review, the establishment of an NSDI as a responsibility of the Department of the Interior was specifically noted:

> By supporting a cross-agency coordinating effort, the federal government can develop a coherent vision for the national spatial data infrastructure (NSDI). (Spatial, or geographic data refers to information that can be placed on a map.) This will allow greatly improved information analysis in a wide range of areas, including the analysis of environmental information and the monitoring of endangered animals and sensitive land areas.

- In a similar vein, the Carnegie Commission on Science, Technology, and Government issued a report[6] that examined the achievements of the states in managing science and technology. The report also recommended ways in which the states can join with industry and the federal government in addressing the domestic issues of the 1990s and beyond: "The establishment of an interstate compact to help the states themselves decide what policies work best in a decentralized and variegated nation."
- The Library of Congress, also exercising a leadership role in this crucial focal area, held a conference on July 14, 1993 on "Delivering Electronic Information in a Knowledge-Based Democracy." The emphasis was on helping shape the policy framework essential to creating an advanced information infrastructure through an examination of the "critical policy issues central to the development of electronic information resources that will be distributed over the emerging digital 'highways.'"
- In September 1993, a State-Federal Technology Partnership Colloquium was "designed to establish broad-scale cooperation between the federal government and the states in matters related to science and technology." Topics such as "telecommunications and information infrastructure," "national and state science and technology policy," and "redefining federal laboratories" were explored.[7]
- In October 1993, a National Research Council report[8] on the National Biological Survey that called for "a new national, multisector, cooperative program of federal, state, and local agencies; museums; academic institutions; and private organizations." A large portion of the proposed programs of the new National Biological Survey involve the generation,

management, and application of biological information tied to geography. The organization will become an important component of the NSDI.
- Several key congressional measures encourage the use of advanced technologies. These include the Stevenson Wydler Technology Innovation Act of 1980, the Technology Transfer Act of 1986, and the Omnibus Trade and Competitiveness Act of 1988. Illustrative of Presidential action (in 1987) augmenting these initiatives was Executive Order 12591, "Facilitating Access to Science and Technology."
- Last but not least, the recent Executive Order (Executive Order 12906, April 11, 1994, "Coordinating Geographic Data Acquisition and Access: National Spatial Data Infrastructure") explicitly calls for the development of partnerships for spatial data acquisition.

Although most of these undertakings have not concentrated on the creation of a national spatial data infrastructure per se, the overarching concerns regarding federal/state cooperative processes, protocols, and priorities are quite germane to the philosophy and programs of those responsible for developing such an infrastructure upon which the nation and our society are increasingly dependent. It is imperative that public and private entities charged with creating the national spatial data infrastructure be fully aware that they represent only a single, but critical, component of a far more comprehensive information infrastructure.

NOTES

[1] Science policy address, February 28, 1993, Palo Alto, California (cited in Executive Summary of the National Association of State Information Resource Executives Case Studies, April 22, 1993, p. 1).

[2] National Association of State Information Resource Executives Case Studies, April 22, 1993.

[3] *National Information and Service Delivery System—A Vision for Restructuring Government in the Information Age*, 1993, State Information Policy Consortium, 8 pp.

[4] *The Information Government: National Agenda for Improving Government Through Information Technology*, 1993, National Academy of Public Administration.

[5] *Creating a Government That Works Better & Costs Less: Report of the National*

Performance Review, 1993, Vice President Al Gore, U.S. Government Printing Office, Washington, D.C., 168 pp.

[6]*Science Technology, and the States in America's Third Century*, 1992, Carnegie Commission on Science, Technology, and Government, New York.

[7]The State-Federal Technology Partnership: Colloquium Proceedings, September 12-14, 1993, Battelle Memorial Institute, Cleveland, Ohio, 38 pp.

[8]*A Biological Survey for the Nation*, 1993, National Research Council, Washington, D.C., 195 pp.

APPENDIX B
COOPERATION/PARTNERSHIP MODELS IN EXISTENCE TODAY

Four examples of multiagency cooperative programs were examined in the course of this study: the U.S. Bureau of the Census, the National Geodetic Reference System, the South Carolina Water Resources Commission, and the Maryland Digital Orthophoto Program. The first two examples have been federally initiated, and the last two are state initiated. These examples provide insight into both the opportunities and the hazards of federal/state partnerships. The degree to which each of the working programs fits a partnership model can be debated; the primary purpose of this review is to identify desirable elements for partnership models.

BUREAU OF THE CENSUS—STATE DATA PROGRAM

The Bureau of the Census is well known as a federal data gathering organization. The public and all state and local agencies are familiar with the agency and its data, and the public has indicated acceptance of the Bureau's work through participation in the data gathering process. To facilitate the data transfer process, the Bureau of the Census has provided for data transfer to the various states through joint statistical agreements. These are formal MOUs that provide for data release, usually through the governor's office. The Bureau of the Census provides the data and the state is obligated to distribute the data and data products within the state. The Bureau provides some limited training, data, and new products to the state. As state computer systems become more sophisticated, the Bureau of the

Census data become more valuable. For instance, Kansas recently underwent a reapportionment based on digital census data. Legislators were able to try their own ideas for district size and configuration on the computer. In consequence, the reapportionment was not challenged in court by a disaffected political party. Additional affiliates of the Bureau of the Census are private concerns, many of whom market value-added products largely based on the original census products.

Businesses constantly use census data for marketing, sales targeting, and demographics. Commercial directories build their businesses on the value adding they do with census data. Costs for the services of this data transfer program are about $800,000 per year, mostly for the 12 full-time staff who distribute the products. All other costs are reimbursed through digital line charges and connect costs. *The primary value of this partnership to the Bureau of the Census is in the visibility of census products. In this way, the bureau is perceived favorably by the public as providing a quality service.*

NATIONAL GEODETIC SURVEY

Geographic information is of little value unless it is tied to the Earth. In the past, the U.S. Coast and Geodetic Survey maintained leveling lines and benchmarks that have been the basis of all land surveying. Public land surveys are also tied to these benchmarks. Now the National Geodetic Survey (NGS) within the National Oceanic and Atmospheric Administration (NOAA) maintains the National Geodetic Reference System and a national geodetic control network. Use of orthophotos and the Global Positioning System (GPS) permits the accurate positioning of all objects on the Earth, but the development of those systems is expensive. The NGS has formed partnerships with transportation, natural resources, and other agencies within the states to provide exchange of survey data and to provide for permanent recording of new survey data. This NGS partnership involves cost sharing. As presented to the MSC, there are several triggers to the development of new surveys, including new highway funding, where federal dollars require the NGS to participate, new technology (GPS), new administrative procedures such as OMB circulars, and various other agency and state needs for survey data to locate objects or sites.

There is a long-standing pride in the NGS deriving from its historical role in the United States and its dedicated clientele and staff. While this program is not as visible as the census program (it does not deal directly with the public), the function is important and should be further studied as a federal/state government model for data partnership.

Mutual planning for surveys and mutual agreement for the use of data as crucial to the success of this program. This is a program for professionals rather than lay citizens. *The primary advantage to the partnership arrangement is the establishment of long-term commitments to create and maintain current data.*

SOUTH CAROLINA WATER RESOURCES COMMISSION

South Carolina is developing a statewide GIS to assist in natural resource management decisions. Federal funding is administered through the South Carolina Water Resources Commission (SCWRC) and NOAA's Strategic Environmental Assessment Division, which are the lead state and federal agencies designated to build this diverse partnership. The project encompasses a broad spectrum of potential GIS user needs and information resources including the local, state, and federal agencies and the private sector. Guidance as to the goals and direction of the project is through an advisory committee that includes the Environmental Protection Agency (EPA), USGS, and NOAA, seven state agencies, several universities, and representatives of the timber and agriculture industry, as well as experts in other fields such as economics, wetland science, anthropology, and landscape architecture.

Initial goals are on placing state-of-the-art information management tools at the disposal of local communities so that natural resource management and economic development decisions are informed and accomplished with minimum conflict. Initial data capture was focused on the nationally significant Edisto River basin where the needs for jobs and for conservation are both critical. Day-to-day operational planning, staff support, and budgetary management are within the SCWRC. Representatives of the SCWRC maintain that the establishment of a lead operational agency has been critical to keeping the project focused and results oriented. They view the strengths of this partnership to be (1) the diversity of the stakeholders, (2) the breadth of funding sources (state, federal, and private sector), and (3) the strict adherence to data quality standards.

One of the first efforts was to establish a user needs study and to design the appropriate land base and natural resource themes. The SCWRC chose the USGS 7.5 minute topographic map base, wetlands, upland land use, soils, environmental permit, and rare species data as the primary conversion targets. This extensive data conversion was financed through joint funding agreements with several federal agencies. As of mid-1993, the USGS and the SCWRC have digitized existing information from about 62 percent of the 353 quadrangles within the state through fifty-fifty cost share agreements. Joint funding agreements are also in place with the U.S. Fish and Wildlife Service to digitize the National Wetland Inventory maps. An in-kind-service agreement exists between the SCWRC and the Soil Conservation Service for quality control so that digital soils information will meet national standards. Additionally, the Army Corps of Engineers and the National Park Service have made significant monetary contributions to the project. Data investment decisions are structured to empower regional and local governments to make informed decisions regarding development and natural resource management. However, data base design decisions have always been influenced by the potential long-term value that these investments will have for water-use planning, emergency response, industry siting, transportation planning, forest products development, and all the various missions that could use the information and technology. Recently, several private businesses with large landholdings in the state have begun negotiations with the SCWRC to add additional funding to the project. The Natural Resources Decision Support System GIS now contains in excess of 20 layers of information and has maintained strict compliance with national map accuracy standards.

Other state and local government agencies are now interested in supporting the system as the data coverage and quality are recognized as appropriate for their use, so that the demand for information and the number of stakeholders continue to increase. *The partnership's success is based on a shared vision that each stakeholder will achieve greater efficiencies in performing their mandates using this information management tool.* Working together will allow each to own and use an application-rich data base that none could afford in isolation. The funding synergy is achieved through a level of mutual respect and codified in joint funding agreements and MOUs. This has been accomplished in spite of outdated rules restricting use of diverse federal funding sources in data conversion matching programs. The investment in building a high quality data base will lead to the development of new applications resulting in even greater benefits to the partnership.

The South Carolina example illustrates a valuable transition to digital data, adding to the resources that have been provided through former

APPENDIX B: PARTNERSHIP MODELS IN EXISTENCE 41

USGS cooperative information exchange and technology transfer programs. The model demonstrates how traditionally independent federal and state agencies can more effectively meet many missions by working together in a focused, well-managed partnership. The model is particularly significant as a statement for the increased benefits that can be derived from increased communication and respect between levels of government. *Finally, as in previous examples, good planning that results in building and maintaining spatial data sets to national standards that meet the needs of multiple stakeholders is the key to measurable benefits and ultimate success of the project.*

MARYLAND DIGITAL ORTHOPHOTO PROGRAM

As the federal government has placed increasing demands on states to develop data and to manage programs without consequent funding, Maryland has developed a cost-effective response that will have wide applicability. The digital orthophoto program (orthophotos are rectified aerial photographs) was developed in response to a mandate to map wetlands, a major issue along the Maryland coast. Previous mapping was not sufficiently accurate for the purpose of this program. The orthophoto data acquisition solved this by creating an image data base that meets accuracy requirements to locate objects or sites the size of individual homes and outbuildings. This not only permitted the wetlands mapping to proceed with great accuracy, it has also generated a great deal of interest in local government for land use planning and property appraisal.

County governments financed part of the data acquisition program by doing the physical monumenting and paying for the aerial photography. The state has funded (using primarily EPA funds) the data reduction and product development. Counties are thus stakeholders, and have been a great source of support for the program in difficult financial times for the state. Major problems for the program, in addition to funding, were concerns of state agencies about the large amount of data resident in a single agency, lack of faith in new technology, and questions of access to data. Simple MOUs and handshake agreements have dispelled these issues, and the system is working.

This model operates like the South Carolina model in one respect: there has been an individual upon whom the system focuses, an agency that leads, and a climate of mutual trust and cooperation. The Maryland system envisions future products to be large-scale small-format distribution system envisions future products to be large-scale small-format distribution of maps, designed on demand for individual citizen use, perhaps by kiosk distribution in shopping malls. While this may or may not reach fruition in the near future, it is clear that vision is required to make federal/state working partnerships effective and to get the job done.

APPENDIX C
EXAMPLE MEMORANDUM
OF UNDERSTANDING

Although many of the proposed partners have examined this possible memorandum of understanding (MOU), it has not been agreed upon and should be taken only as an example. This example used Ohio, however, the general content could be similar elsewhere. The proposed partners in this example are the U.S. Soil Conservation Service (state office), the Ohio State Department of Natural Resources, the Ohio State Environmental Protection Agency, and county auditors. Again none of these organizations have formally endorsed this example.

INTRODUCTION

A current effort within the Soil Conservation Service (SCS) is to digitize the original soil survey maps for the Soil Survey Geographic Data base (SSURGO). SSURGO is the most detailed level of soil mapping done by the SCS. This level of mapping is designed for use by landowners, townships, and county natural resource planning and management.

The mapping bases used meet national map accuracy standards and are either orthophotoquads or 7.5-minute quadrangles. SSURGO is linked to a Soil Interpretations Record attribute relational data base, which includes over 25 soil, physical and chemical properties for approximately 18,000 soil series recognized in the United States. Information that can be queried from the data base includes available water capacity, soil reaction, salinity, flooding, water table, bedrock, and interpretations for septic tank limitations, engineering, cropland, woodland, and recreation development.

SSURGO data are available in either the USGS Digital Line Graph (DLG-3) Optional Distribution Format or the SCS Geographic Exchange Format. SCS soil map symbols (e.g., AbC) are made available as ASCII file when SCS soils data are distributed in the DLG format.

NEED FOR DIGITAL SOILS DATA

The broad value of digital soils data is widely recognized at the federal, state, county, and local levels. Four agencies are identified in this agreement with a common need for soils data: the SCS, the State Department of Natural Resources (SDNR), the State Environmental Protection Agency (SEPA), and the County Auditors (Counties), representing county government.

• The SCS, in carrying out its responsibilities in the National Cooperative Soil Survey, has need for digitized soil survey data. The SCS uses these data for conservation planning, watershed management, engineering interpretations, and to help prioritize conservation and land use needs.

• The SDNR in carrying out its assigned responsibilities under applicable state law has need for soil and water resources data from conservation programs to meet identified state and local needs.

• The SEPA uses soils data to improve their ability to evaluate ground water pollution, aquifer and wellhead protection; perform best management practice selection and watershed prioritization; and develop pollutant-loading models and soil erosion models.

• The Counties use soils data for purposes such as siting decisions, to determine soil limitations for various uses, and to determine agricultural land valuations on parcels.

CURRENT STATUS

Since 1987 through a cooperative agreement, the SDNR and the SCS have jointly produced SSURGO data at a rate of three to four counties per year. At this rate it will take 20 years to complete SSURGO coverage of

APPENDIX C: EXAMPLE MOU

the state. Where SSURGO data are not available, the Counties, SEPA, and other groups often commission private firms to digitize the data or do it themselves. Usually soils data generated by one of these groups cannot be used by others, due to lack of common standards for data format and quality, and the limited area of coverage. For these reasons and to meet SCS standards and specifications, the data must also be digitized later into the SSURGO format. Over time, this approach results in a costly duplication of effort; therefore, an accelerated schedule for generating SSURGO data is highly desirable.

TERMS OF AGREEMENT

In consideration of the mutual benefits to be derived, the SCS, the SDNR, and the SEPA agree to the following:

1. To accelerate the digitation of soil surveys in the state by targeting a five-year production schedule to complete coverage of the entire state, contingent upon availability of suitable base maps, soil recompilations, and funding.

2. To complete the digitizing and delivery of output products for mutually selected counties using standards specified in Technical Specifications for *Line-Segment Digitizing Soil Survey Maps*. The use of SCS standards and specifications to digitize soils surveys of the second order will result in a data format usable by all parties in this agreement, other federal agencies, state, and local agencies and the public.

3. That the State Department of Natural Resources will be the lead agency in carrying out this agreement, under the coordination of the Department GIS Coordinator.

4. That funding and/or cost sharing will be furnished according to this formula:

50% of cost	SCS
16.67% of cost	SDNR
16.67% of cost	SEPA
16.67% of cost	Counties

Support will take the form of cash to SDNR, and/or cost-sharing in the form of contributed labor, hardware, software, or other means agreed by all parties in a written contract. SDNR may, at its discretion, perform the work in accordance with the funding provided, or subcontract the work.

5. That the SDNR will maintain a permanent archive of SSURGO data for the state and distribute them to state residents, business, and agencies, in accordance with state law.

6. That the SCS will also maintain a permanent archive of SSURGO data and distribute them according to its standard pricing schedule.

7. That SCS, SEPA, SDNR and Counties will each appoint a representative to serve on a coordinating committee. Each representative will coordinate activities and funding on behalf of his/her agency. The SDNR representative will be the Department GIS Coordinator, who will chair the committee. Each representative will have equal decision-making authority. In addition, representatives will assist in planning the work, preparing agendas, participating in public meetings if necessary, and other similar activities.

8. That SCS, SEPA, SDNR and Counties intend to maintain the soils data in a long-term partnership arrangement. Updates to the soil maps will be digitized and funded according to the terms of this agreement.

It is mutually agreed that this agreement may be terminated if either SCS, SDNR, and SEPA fail to comply with any of the conditions of this agreement.

APPENDIX D
STATE GEOGRAPHIC INFORMATION AUTHORIZATIONS AND COORDINATION—SUMMARY

At the end of this appendix is a comprehensive listing of 100 state directives, including statutes, executive orders (by governors) and memoranda of understandings (MOUs) which mention or directly influence geographic information. All three are considered "authorizations," though only the first two are truly directives. While executive orders and MOUs can be as effective as statutes for coordination, statutes have a longer duration (beyond individual governors and signatories to agreements) and are generally needed to provide funding and authority to ensure commonality, compliance, and oversight.

Almost half (49) of these directives were authorized during or after 1991. At least five additional directives (probably executive orders) are expected to be authorized before the end of 1993 but this listing only included through September 1993. While the various state directives differ, below is a general description and findings for each type.

STATE STATUTES

Over two-thirds of the states have at least one statutory reference to geographic information in one form or another, with some, such as Louisiana, Minnesota, and North Carolina having multiple references. The compilation can be considered complete as of the end of most 1993 legislative sessions. Statutes address the following (followed by the number of states having each type):

1. Authorize geographic information coordination groups or studies (17, including two groups that are no longer in existence.);
2. Authorize statewide or broad environmental geographic information offices, data bases, or funding (14);
3. Direct geographic information use or data development for specific missions or needs, mainly natural resources management, environmental protection or growth management (11);
4. Provide for access and cost recovery for geographic (spatial) data, often modifying open records laws and directly impacting localities (10); and
5. Other matters, including providing some help to local and regional entities, reapportionment use, requiring compatibility of state-funded data (Minnesota), and directing the private sector to develop compatible data (New Jersey).

While there is an increase in statutory references in all of the above categories, few omnibus statutes exist specifically for geographic information, few authorize offices or funding accordingly, and few have "teeth" to require commonality or oversight. The most omnibus state statutes are in Maine and Utah, both adopted in 1991 and establishing offices along with other geographic information direction, with Minnesota adopting legislation to officially authorize its 15-year-old Land Management Information Center in 1992.

EXECUTIVE ORDERS

All of the 15 reported executive orders were signed to establish councils, set direction for member state geographic information coordination groups, and possibly to establish statewide geographic information centers, as in Idaho, Kentucky, North Carolina, and Oregon. Two of the orders are no longer in effect. Four states have pending executive orders to establish or sanction existing geographic information groups.

MEMORANDA OF UNDERSTANDING

Five MOUs are reported. Four MOUs are to promote general geographic information coordination, with two the primary authorizing instrument for a geographic information group (Colorado and Montana). North Carolina's MOU is specifically for GPS and complements executive order direction. While not reported in the compilation, many state MOUs exist that describe geographic information relationships between two or more agencies, for example for geographic information services, projects, and data exchange.

DISCUSSION

These authorizations typify the overall findings described above. geographic information coordination is evolving and has growing support at statewide levels, as evidenced by official sanctions for coordination and state geographic information centers. Many of these directives provide for increased multisectoral participation. However, almost in contrast, some of these directives are organizationally placing geographic information direction under broad information policy and organizations that generally only address state government. To date only a few directives establish any oversight, or require data commonality and compliance.

The compilation includes a list, description, and identification of membership of statewide geographic information coordination groups. While there has been a significant increase in regional interlocal groups at the substate level, none are included here. It also includes at least one group in each of the 50 states. All but two groups (in Alabama and Delaware, though Delaware has another group) are considered active at this time. Virtually all of the groups reported are multiagency in focus and membership. Of the 21 states showing more than one group, seven have one of the reported groups officially reporting to another reported group. Many of the reported groups have subgroups that are not identified here.

Overall, there is a growing trend toward a broad naming and focusing of state geographic information coordination groups to consider all geographic information and related technologies. However, 26 states have groups with the term "GIS" in their name. Some of these groups have a limited view, focusing specifically on GIS, while others are broader. It appears that a generally accepted and emerging name in some (15) states is "Geographic Information Council" or "Committee," as is reflected by the naming of the National States Geographic Information Council (NSGIC). Some states have both a group with geographic information in the name and another with GIS. Focus on geographic information rather than GIS is to represent a broad view and encourage attention to data rather than technology. Certain other words are also commonly used in group names, such as "land" (5 states).

Nine states have a State Mapping Advisory Committee (SMAC) reported to be a statewide geographic information coordination group. Of these, only the SMACs in Nevada, New Jersey, and Oregon are broad in focus and exist without other, more influential geographic information groups in their state. These three are in effect geographic information councils, and SMACs in name only. Most states now have SMAC functions under broader, higher-level groups.

The groups represent a full range of authority and levels of attention to both policy and technical matters. Over 40 states have at least one group with some degree of official stature, via statute (11), executive order (14), memoranda of understanding (2), or other method, such as by agency leaders. A wide range exists in terms of level of policy voice and technical issues addressed by groups. Both policy and technical issues can be addressed by the same group, particularly if they are organized under general information groups, or by two groups, with one reporting to another. Some states have what are essentially GIS users groups, with or without other groups.

Membership is the focus of the remainder of this discussion about the groups. However, not reflected in the table is the organizational level of the individual participants in the groups. Overall, these levels mirror general geographic information conditions in states, and the wide range of differences in attention to policy verses technical issues. Members can range from agency directors (as in Kentucky and North Carolina) to mid-level or policy-level agency officials or, at the other end of the scale, GIS users. Membership representation in the groups is identified by sectors and state governance functions.

The survey lists all of the sectors identified as members of state geographic information coordination groups. All groups have state government representation. All but ten groups are multisectoral, with academia and then local government being the most often mentioned membership sectors. Participation by the other sectors can vary significantly. For example, there can be many individuals for a sector such as in Montana where as many federal agencies as state agencies are involved. In other states, such as Washington, one federal official is essentially representing the entire federal government. The participation issue is also a problem regarding localities. Some states have individual local representatives which may or may not represent all localities. The most noteworthy under-represented sectors may be Native Americans and utilities, though they are very important for a national approach.

The survey lists virtually all of the state government functions identified as members of state geographic information coordination groups. An overall classification of governance functions and specific state entities identified are as follows:

1. **Branches** (legislative, judicial, executive)
2. **General government and administration** (Governor, Planning; Budget, Finance and Comptroller; Secretary of State; Administration; Personnel, Human Development; Revenue, Property Tax Administration; Insurance, Regulation)
3. **General information** (Information Policy, Information Statistics, Library, Information Technology, Census Data Center, Archives, Records Management, State Surveyor, State Cartographer)
4. **Natural resources management and environmental protection** (also Public Land Management, State Forestry, State Geological Survey; Cultural Resources, Archeological Survey, Historical Preservation, Agriculture)
5. **Infrastructure** (Transportation, Regulatory Utilities Commission)
6. **Human/Social Services** (Social Services, Aging, Youth Programs; Human Health; Employment Security, Labor; Education; Higher Education)
7. **Public Safety, Emergency Management**
8. **Economic Development/Growth Management** (Economic Development, Commerce, Tourism; Rural Development, Community, Local Affairs)

From the compilation, it can be concluded that virtually all functions of governance are increasingly represented in the geographic information coordination groups. This trend is important and positive in terms of developing and implementing omnibus efforts on a statewide basis. This condition also directly impacts coordination opportunities with the federal government. Of these functions, the primary one is natural resources management and environmental protection, represented on virtually all state groups. Transportation is included in most as well. Representation of each of the other functions is increasing, particularly during the last three years.

ALABAMA

AUTHORIZATION FOR GEOGRAPHIC INFORMATION/GIS

Citation	Name/Type	Description
Spring 1991	Memorandum of agreement for the coordination of GIS and land information systems among universities	Signed by representatives of ten of the state's major universities to help initiate and build a statewide university level program that combines the strengths of each university.

GEOGRAPHIC INFORMATION/GIS COORDINATION GROUP

GIS/LIS Information Exchange Group (unofficial)

Mission: The group's objective is to be a "grass roots" coordination body to reflect statewide GIS needs to help guide management toward achievable goals.

Sectors Represented: State, federal, local, regional entities, utilities, academic, professional organization, private sector users, private sector suppliers

Functions: Environmental protection-natural resource management, employment-labor, economic development, legislative, state geological survey, higher education

ALASKA

AUTHORIZATION FOR GEOGRAPHIC INFORMATION/GIS

Citation	Name/Type	Description
AS 09.25.110,115 AS 14.56.120(b) (1990)	Requests for information by public agencies, public access	The law was enacted to authorize sales of electronic products and services and to provide a foundation for citizen access to GIS technology. It distinguishes between "public records" and "electronic services and products."

GEOGRAPHIC INFORMATION/GIS COORDINATION GROUPS

Committee on Natural Resource Information Management (CONRIM) (official via MOAs)

Mission: CONRIM is organized for the purpose of improving the acquisition, management, and dissemination of information among government, academic, research, and private interests concerning Alaska and the Arctic by promoting networking of data, technology, personnel, and resources related to information management.

Sectors Represented: State, federal, local, academic, tribal, private sector users

Functions: Governor's office, legislative, budget-comptroller, environmental protection-natural resource management, public lands management, forestry, state geological survey, cultural resources-archeological-historical, agriculture, transportation, health

State Mapping Advisory Committee (official)

Mission: The SMAC was organized to coordinate state requirements for mapping and communicate them to USGS. SMAC's goals include exchanging information on data holdings, systems, needs and plans; identify statewide map and data requirements; establish data exchange standards; plan for data; and coordinate.

Sectors Represented: State, local, tribal, federal, academic

Functions: Ggovernor's office, state planning, information technology, community-local affairs, state surveyor, environmental protection-natural resource management, public lands management, forestry, state geological survey, cultural resources-archeological-historical, agriculture, transportation, economic development, rural development

ARIZONA

AUTHORIZATION FOR GEOGRAPHIC INFORMATION/GIS

Citation	Name/Type	Description
Executive Order 89-24	Establishes Arizona Geographic Information Council	Established AGIC and terminated the Arizona State Mapping Advisory Committee. AGIC serves as an advisory council to the Arizona State Land Department. Roles include providing guidance and direction in the management of state GIS.
Executive Order 92-17	Membership of AGIC expanded	Expands the membership of AGIC to include the state's four regional GIS consortia.
ARS Ch. 37-172,173 amended 1984 & 1988	Resource Analysis Division	This chapter originally established the Resource Analysis Division in the State Land Department, and in 1988 established the Office of the State Cartographer in the Department.
ARS Ch. 37-176 (1992)	Resource Analysis Division (update)	Established nonlapsing, dedicated revolving fund to support GIS activities, source of monies, claims

GEOGRAPHIC INFORMATION/GIS COORDINATION GROUP

Geographic Information Council (official via EO)
Mission: The mission of AGIC is to coordinate the development and management of statewide geographic information and serve as an advisory council to the Arizona State Land Department to provide guidance and direction in the management of a state GIS, including standards for data architecture, quality, etc.
Sectors Represented: State, federal, local government via association, academic, regional entities, private sector users
Functions: V - revenue-property tax, environmental protection-natural resource management, transportation, employment-labor, education, economic development, public lands management, health, administrative, state geological survey, higher education;
NV - social services, forestry, state surveyor, census data center, information technology, personnel-human development

ARKANSAS

AUTHORIZATION FOR GEOGRAPHIC INFORMATION/GIS

Citation	Name/Type	Description
Act 150 (1993)	Creates the State Mapping and Land Records Modernization Advisory Board	Creates the board, designates it as the State Mapping Advisory Committee to coordinate with USGS, provides that the board will ensure that digital map data meets or exceeds national standards, encourage coordination and eliminate duplication, and to provide recommendations for a statewide program.

GEOGRAPHIC INFORMATION/GIS COORDINATION GROUPS

GIS Users Forum (official by state leader)

Mission: The objective of the forum includes development of goals, identification of major GIS issues to encourage data sharing and improve data compatibility, development of standards, and determination of appropriate legislation.
Sectors Represented: State, federal, local, utilities, academic
Functions: Governor's office, census data center, environmental protection-natural resource management, transportation, regulatory utilities commission, cultural resources-archeological-historical, employment-labor, economic development, public safety-emergency management

Mapping and Land Information Modernization Advisory Board (official by statute)

Mission: The mission of the board is to study the current mapping and land information situation and identify goals and objectives including standards, encourage coordination, minimize duplication. Board will dissolve after providing recommendations to the governor, probably to include additional legislation.
Sectors Represented: V - state, utilities, professional organizations, local government via association, academic; NV - federal, professional organization, academic
Functions: V - legislative, information policy, information technology, environmental protection-natural resource management, transportation, rural development, cultural resources-archeological-historical; NV - health, public safety-emergency management, state surveyor

CALIFORNIA

AUTHORIZATION FOR GEOGRAPHIC INFORMATION/GIS

Citation	Name/Type	Description
Public Resources Code, Sec.8900 (1991)	Creates of Geographic Information Task Force	The Geographic Information Task Force was directed to provide recommendations to the governor and legislature concerning task force mechanisms to fund, develop, exchange, and maintain geographic information among public and private sector entities in the state.
Executive Order W-22-92	Establishes members, chair, participants, and staff of GI Task Force	The executive order follows the statute creating the GI Task Force by establishing the chair to be the director of the Governor's Office of Planning and Research, members to include leaders of five state agencies, participants to include representatives of academia, USGS, BLM, USFS, NOAA, and associations, and staff by Teale Data Center.

GEOGRAPHIC INFORMATION/GIS COORDINATION GROUP

Computer Mapping Coordinating Committee (unofficial)

Mission: The committee serves as a users group and a forum for the development of policy statements and guidelines regarding the use of GIS and related data. It also assists in the development of standards and specifications for digital data and strives to serve as a clearinghouse.

Sectors Represented: State, academic, federal, regional entities, local

Functions: Information technology, environmental protection-natural resource management, transportation, state geological survey, forestry, public lands management

COLORADO

AUTHORIZATION FOR GEOGRAPHIC INFORMATION/GIS

Citation	Name/Type	Description
D015089 October 19, 1989	Executive order: Colorado Geographic Information Coordinating Committee	The executive order established the CGICC to promote co-operation between state, federal, and local agencies, and the private sector in addressing geographic data and information needs and services. (The executive order had a sunset date of the end of 1991—it was supplanted by theMOU)
Spring 1993	Memorandum of understanding for interagency GI coordination and GI Coordination Committee	The MOU was initiated and signed by seven state agencies and the Legislative Council to document agreement to shared objectives regarding GI/GIS data, technology, and investments, including maximizing and facilitating opportunities. It establishes the membership, responsibilities, and procedures of the GICC.

GEOGRAPHIC INFORMATION/GIS COORDINATION GROUP

Geographic Information Coordinating Committee (official by MOU)

Mission: The role of the committee is to promote cooperation between state, federal and local agencies, and the private sector as a means to strive for excellence in the management of state resources. It also acts as an educational forum for GI in Colorado.
Sectors Represented: **V** - state, local government via association; **NV** - federal
Functions: Administrative, environmental protection-natural resource management, transportation, health, community-local affairs, agriculture, public safety-emergency management, legislative

CONNECTICUT

AUTHORIZATION FOR GEOGRAPHIC INFORMATION/GIS

Citation	Name/Type	Description
Public Act 91-249 (1991)	Authorizes that municipalities may charge a fee for use of GIS	This one-sentence act states "any municipality may by ordinance impose a reasonable fee for the use of its GIS."

GEOGRAPHIC INFORMATION/GIS COORDINATION GROUPS

GIS Committee (unofficial)
Mission: This committee is a volunteer group interested in promoting and developing land information systems, including land records modernization. It is working with the GIS Policy Committee.
Sectors Represented: State, local, professional organization, private sector suppliers, academic, private sector users
Functions: transportation, environmental protection-natural resource management, information policy

GIS Policy Committee (official)
Mission: The role of the committee is to investigate issues in state government and then address the state's role beyond state government concerning geographic information. The group is preparing findings and recommendations accordingly.
Sectors Represented: Sstate, academic
Functions: Sstate planning, information policy, information technology, environmental protection-natural resource management, transportation, economic development

DELAWARE

AUTHORIZATION FOR GEOGRAPHIC INFORMATION/GIS

Citation	Name/Type	Description
1991 H.J.R. 17	GIS Oversight Committee	Establishes a GIS oversight committee to coordinate GIS activities and resources; however, this committee was never formalized or organized (other committees pending).

GEOGRAPHIC INFORMATION/GIS COORDINATION GROUPS

GIS Committee (official via statute)
Mission: The committee was established to coordinate GIS activities and resources, but it never was constituted.

Statewide GIS Committee (unofficial)
Mission: The mission of the committee is to coordinate GIS activities.
Sectors Represented: State
Functions: Environmental protection-natural resource management, transportation, agriculture, public safety-emergency management, state geological survey, health

FLORIDA

AUTHORIZATION FOR GEOGRAPHIC INFORMATION/GIS

Citation	Name/Type	Description
Florida Statutes 282.403	Creates Growth Management Data Network Coordinating Council	Creates the Growth Management Data Network Coordinating Council to facilitate the sharing of growth management information. In 1988 a strategic plan focused the council's activities on geographic information and GIS.
Florida Statutes 253.023(4)	Land inventory for land acquisition	Initiate and maintain a Natural Areas Inventory to aide in the identification of areas to be acquired in the Conservation and Recreation Lands Trust Fund.
Florida Statutes 253.0325	State Lands Records Modernization	This program provides for an on going computerized information systems program to modernize its state land record and documents that relate to land to which title is vested.

GEOGRAPHIC INFORMATION/GIS COORDINATION GROUPS

Base Mapping Advisory Committee (official by EO)

Mission: The BMAC was organized to serve as a technical advisory committee for both manual and digital mapping. Its focus includes developing a geodetic base map, multipurpose planimetric maps, and a multipurpose cadastral (parcel) map that can be used for local property tax and appraisal.
Sectors Represented: State, regional entities, local
Functions: Environmental protection-natural resource management, transportation

Data Network Technical Advisory Committee (official via council)

Mission: This committee serves as "subcommittees" of the Council as formed upon the recommendation of council members, Staff Advisory Committee, or others. These subcommittees are issue specific, and membership varies by subcommittee.
Sectors Represented: State, local, local government via association, professional organization, private sector suppliers, nonprofit, media, academic, utilities
Functions: State planning, revenue-property tax, environmental protection-natural resource management, transportation, agriculture, health, public safety-emergency management, legislative, information technology, state surveyor

Data Network Staff Advisory Committee (official via council)
Mission: The role of the Staff Advisory Committee is to prepare recommendations and conduct work for the Growth Management Data Network Coordinating Council, while serving as a liaison to the council members, council staff, and the technical advisory committee.
Sectors Represented: **V** - state
Functions: State planning, revenue-property tax, environmental protection-natural resource management, transportation, agriculture, health, public safety-emergency management, legislative, information technology, state surveyor

Growth Management Data Network Coordinating Council (official via statute)
Mission: The mission of the council is to lead the coordination of growth management information, recognizing that all such information can be referenced by location. The council's goals include providing growth management direction and guidance to state, regional, and local agencies accordingly.
Sectors Represented: **V** - state; **NV** - local, regional entities, private sector users
Functions: **V** - state planning, revenue-property tax, environmental protection-natural resource management, transportation, public lands management, agriculture, economic development, community-local affairs, health; **NV** - information policy, information technology

GEORGIA

AUTHORIZATION FOR GEOGRAPHIC INFORMATION/GIS

Citation	Name/Type	Description
Senate Bill 437 (1988)	To create the State Mapping and Land Records Modernization Advisory Board	The board was established to study land records modernization and is administered by the Department of Community Affairs.

GEOGRAPHIC INFORMATION/GIS COORDINATION GROUP

State Mapping and Land Record Modernization Advisory Board (official via statute)

Mission: The mission of the board is to study land records modernization and help implement land information systems in local governments through financial incentives, education, and standards and technical specifications.

Sectors Represented: State, regional entities, local, professional organization

Functions: Legislative, secretary of state, revenue-property tax, environmental protection-natural resource management, transportation, community-local affairs

HAWAII

AUTHORIZATION FOR GEOGRAPHIC INFORMATION/GIS

Citation	Name/Type	Description
Hawaii Act 103 (1993)	Authorizes state and local agencies to charge fees for GIS digital data	This act authorizes state and local agencies to charge fees for GIS digital data. It provides that the cost of reproducing GIS digital data "shall be in accordance with rules adopted by the agency having charge or control of that data" and can include labor cost, materials, electricity, equipment, etc.
Hawaii House Resolution 275 (1987)	Relating to GIS	It resolved that the Legislature ask the Department of Business and Economic Development to chair a GIS Task Force, later changed to the new Office of State Planning in the Governor's Office.

GEOGRAPHIC INFORMATION/GIS COORDINATION GROUP

GIS Task Force (official via statute)
Mission: The mission of the GIS Task Force has been to define responsibilities among state agencies participating in statewide GIS activities. The task force has worked to develop an implementation strategy and build statewide data bases for use with GIS, including appropriate standards.
Sectors Represented: State
Functions: State planning, budget-comptroller, economic development, agriculture, public safety-emergency management, public lands management, environmental protection-natural resource management, transportation, health, information policy, information technology

IDAHO

AUTHORIZATION FOR GEOGRAPHIC INFORMATION/GIS

Citation	Name/Type	Description
Executive Order 92-24	Reauthorizes the GI Advisory Committee and GI Center	This executive order continues the GI Advisory Committee created by EO in 1988 and preceding EOs, and establishes memberships and responsibilities. The order also reaffirms that the Department of Water Resources manages the newly named Idaho GI Center, replacing past orders, and defines its coordination and clearinghouse roles.
Idaho Code 39-102 (1989)	Ground Water Protection Act	The Department of Water Resources has the responsibility to maintain the natural resource GIS for the state and is the collector of baseline data for the state's water resources.

GEOGRAPHIC INFORMATION/GIS COORDINATION GROUP

Geographic Information Advisory Committee (official via EO)

Mission: IGIAC's role is to coordinate geographic information, including GIS, remote sensing, and digital and manual cartography, of various sectors of the state. Efforts are coordinated with general information policy and include data standards and specifications for GIS technology.

Sectors Represented: V - state; NV - federal, academic, private sector users, utilities, tribal, local

Functions: Budget-comptroller, revenue-property tax, environmental protection-natural resource management, public lands management, transportation, health, social services, higher education, administrative, forestry, state geological survey, cultural resources-archeological-historical

ILLINOIS

AUTHORIZATION FOR GEOGRAPHIC INFORMATION/GIS

Citation	Name/Type	Description
House Joint Resolution 1 (1993)	Creates a Task Force on Geographic Information Management Technology	This resolution creates a Task Force on GI Management Technology, including defining its membership and roles. It is directed to consider interests of various sectors and develop a strategic plan in 1994.
Executive Order 10 (1991)	Revises the Governor's Science Advisory Committee to include environmental and economic database	This order expanded the Governor's Science Advisory Committee into the environmental arena and increased its importance in state activities. It also directed the Committee to work with the Dept. of Energy and Natural Resources and others to produce a biennial report on the State of Illinois' environment (using GI/GIS).
Ill. Rev. Stat. Ch. III, §7056 (1988)	Solid Waste Management Act	Application of Department of Energy and Natural Resources GIS for solid waste.

GEOGRAPHIC INFORMATION/GIS COORDINATION GROUPS

(ENR) Illinois GIS Policy Committee (official via agency director)
Mission: The role of the IGIS Policy Committee is to coordinate policy issues and activities related to the Department of Energy and Natural Resources GIS, including budget, finance, and decisions about system use, expansion, and maintenance.
Sectors Represented: State
Functions: Environmental protection-natural resource management

Geologic Mapping Advisory Committee (official by agency)
Mission: The mission of the committee is to set state priorities for geologic mapping, in response to the need for more of such mapping and the new national program to help fund and accomplish it.
Sectors Represented: State, federal, local, academic, private sector -users, private sector -suppliers
Functions: Environmental protection-natural resource management

Mapping Advisory Committee (unofficial)
Mission: The role of IMAC is to serve as a single voice for all state map users to make recommendations to USGS about mapping priorities and requirements. It also provides an opportunity for information exchange and worked to accelerate topographic mapping production.
Sectors Represented: State, federal, local, academic, private sector users, private sector suppliers
Functions: Environmental protection-natural resource management

Task Force on GI Management Technology (official by legislation)
Mission: The Task Force on GI Management Technology was formed to review existing and forecasted developments in Illinois, and formulate a strategic plan for development and coordination of this technology.
Sectors Represented: State, academic
Functions: Information technology, legislative, environmental protection-natural resource management, revenue-property tax, agriculture, economic development, transportation, regulatory utilities commission, community-local affairs

INDIANA

AUTHORIZATION FOR GEOGRAPHIC INFORMATION/GIS

None

GEOGRAPHIC INFORMATION/GIS COORDINATION GROUPS

State GIS Forum (unofficial)

Mission: The forum's mission is to identify state, federal and local agency GIS needs, assist in standards development for acquisition, implementation, utilization and maintenance of the technology, identify resource needs and data sharing opportunities, and educate state staff and others regarding GIS.
Sectors Represented: State, federal, local, academic, private sector users, private sector suppliers
Functions: Environmental protection-natural resource management, public safety-emergency management, forestry, state geological survey, budget-comptroller, information policy, library, administrative, information technology, census data center, transportation, health, social services

University GIS Alliance (unofficial)

Mission: The alliance was formed to establish a cooperative way for the state's universities to assist governments and the private sector with all aspects of GIS. It was organized to identify and capitalize on the strengths of each university and hold educational programs in coordination with government.
Sectors Represented: Academic
Functions: -

IOWA

AUTHORIZATION FOR GEOGRAPHIC INFORMATION/GIS

Citation	Name/Type	Description
Iowa Code 22.2 3 (1989)	Act relating to the establishment and financing of geographic data se systems by cities and counties	This section of the statute states that a governmental body that maintains a geographic computer data base is not required to permit access to or use of the data base except according to terms of the governing body. The governing body shall establish rates and procedures for the retrieval of records.
Iowa Code 17 455E.8 (1988)	Groundwater Protection Act	It directed that the Department of Natural Resources shall, develop and maintain a natural resource GIS and comprehensive water resource data system.

GEOGRAPHIC INFORMATION/GIS COORDINATION GROUP

GI Committee (unofficial)

Mission: The mission of the committee is to provide a forum for information sharing about GI/GIS among state agencies.

Sectors Represented: State

Functions: Library, census data center, environmental protection-natural resource management, public lands management, forestry, state geological survey, cultural resources-archeological-historical, agriculture, transportation, regulatory utilities commission, health, education, public safety-emergency management, economic development

KANSAS

AUTHORIZATION FOR GEOGRAPHIC INFORMATION/GIS

Citation	Name/Type	Description
Senate Bill 793 (1990)	Appropriation for common land data base	The legislature appropriated $500,000 to the GIS Policy Board for development of a common land data base for use by 21 federal, state and local agencies.
Kansas SA 74-7701 (1984)	Established the Kansas Commission on Applied Remote Sensing	The statute stipulates that the commission shall assist users, act as a forum for interagency coordination, advise the KARS Program, and disseminates information about remote sensing and GIS.
Executive Directive of March 14, 1989	Establishes the state GIS Initiative including the GIS Policy Board	This directive of the governor establishes the Kansas GIS Initiative and its objectives, including access, coordination, compatibility, sharing, enhanced analysis and decision making, and reduced costs. The membership of the Policy Board and its roles are included, such as establishing policies, data priorities, etc.

GEOGRAPHIC INFORMATION/GIS COORDINATION GROUPS

GIS Policy Board (official by governor)

Mission: The GIS Policy Board's role is to address policy issues related to GIS development and management, including data access, data base priorities, standards, funding and interagency coordination. The board works for compatibility of GIS technology among agencies and leads statewide GIS.

Sectors Represented: State, local, federal, academic

Functions: Revenue-property tax, environmental protection-natural resource management, transportation, health, governor's office, budget-comptroller, information policy, information technology, state geological survey, secretary of state, legislative, administrative, social services, public lands management, forestry, agriculture, regulatory utilities commission, economic development

Technical Advisory Committee (official by Policy Board)
Mission: The committee's mission is to address technical issues related to GIS development and implementation, and to provide input to the GIS Policy Board.
Sectors Represented: State, local, federal, professional organization, private sector users, private sector suppliers, academic, utilities
Functions: Legislative, information technology, information policy, environmental protection-natural resource management, health, secretary of state, agriculture, regulatory utilities commission, economic development, revenue-property tax, transportation, budget-comptroller, administrative, public lands management, forestry, state geological survey, social services, public safety-emergency management

KENTUCKY

AUTHORIZATION FOR GEOGRAPHIC INFORMATION/GIS

Citation	Name/Type	Description
Executive Order 92-1049	Establishes the GI Advisory Council and creates the Office of GI	The order provides that the council be formed and report to the Information Systems Commission. It also defines its membership and duties, including planning and assessing state agency compliance with GIS standards through the information planning process. The Office of GI provides assistance, clearinghouse services, etc.
KRS Sec. 61.970, 975 (1992)	Open records (public access to government data bases)	Directed specifically at GIS data and GIS; exempts GIS data base from public disclosure if for commercial purpose, fees for copying GIS, even if not for commercial purpose if nonstandard "product," penalty for abuse of commercial use.
KRS s 5.010, Title 2 & 118B.010, 10, 1991	Redistricting use of maps	Regarding establishment of districts for the Legislature and the U.S. House of Representatives and the source of maps for such purpose.
KRS s 176.435, Title 15, Ch. 176	Use of electronic data bases on highway construction programs	Influenced use of electronic data bases on highway construction program.

GEOGRAPHIC INFORMATION/GIS COORDINATION GROUP

GI Advisory Council (official by EO)

Mission: The GI Advisory Council was established to assist state and local jurisdictions in developing, deploying and leveraging GI resources to improve public service administration. Roles include developing a GIS plan and model, including data priorities, assess agency compliance with GIS standards

Sectors Represented: State, local, academic

Functions: V - transportation, social services, state geological survey, revenue-property tax, information policy, information technology, economic development, community-local affairs, public safety-emergency management, education, environmental protection-natural resource management, agriculture, insurance regulation; NV - legislative

LOUISIANA

AUTHORIZATION FOR GEOGRAPHIC INFORMATION/GIS

Citation	Name/Type	Description
Revised Statute 50:171, 1989	Establishes standards for a Statewide Land Information Mapping and Map Records System	The statute requires the Department of Natural Resources, Office of State Lands (later transferred to the Division of Administration) "establish, promulgate, and maintain appropriate standards for a statewide land information mapping and map records system."
R.S. 39:291-298 (1991) (Act 728)	Establishes State Data Base Commission	This act creates the Louisiana Data Base Commission, which shall be responsible for the creation and maintenance of a data base for the state. The purpose of the data base is for strategic planning, policy formulation, and administration of state government. (It does not mention GIS, but efforts are coordinated with GIS work.)
MOU December 1, 1989 and amended	MOU including state agencies, legislature, and attorney general	The purpose is to establish "an efficient and effective method for the coordination, research and development, access to, and technical assistance of geographically related information within state government."
Revised Statute 50:172-174, 1991	Relative to land information and records, creating Land Information Advisory Board, standards, and parish offices	This statute provides that the Division of Administration's Office of State Lands shall provide standards, technical assistance and advice to local governmental units and assessors. A Land Information Advisory Board is created to review and recommend standards.
Senate Current Res. No. 44, 1991	To create the GIS Study Commission	This resolution creates the GIS Study Commission to study and make recommendations relative to GIS to the legislature.
House Concurrent Res. No. 171, 1990	To recognize the Task Force for a Louisiana GIS Network	This resolution was adopted to recognize the task force as a Louisiana GIS network and to "urge and request the division of administration to attempt to secure federal funds for the purpose of said task force."

GEOGRAPHIC INFORMATION/GIS COORDINATION GROUPS

Data Base Commission (official by statute)

Mission: The mission of the Data Base Commission is to establish a state data base for strategic planning, policy formulation, and administration of state government.

Sectors Represented: Sstate, private sector suppliers, academic

Functions: Administrative, legislative, and rotates agriculture, judicial, insurance regulation, elections, archives-records management, treasurer

Task Force for a Statewide GIS Network (official by legislation, MOU)

Mission: The task force's mission includes development of a strategic plan and uniform standards for GIS and geographic information, including a needs analysis and cost/benefit study to determine the needs of the state including local governments.

Sectors Represented: State

Functions: Legislative, judicial, administrative, information technology, state planning, public lands management, environmental protection-natural resource management, economic development, transportation, agriculture, health, public safety-emergency management, cultural resources-archeological-historical

MAINE

AUTHORIZATION FOR GEOGRAPHIC INFORMATION/GIS

Citation	Name/Type	Description
12 MRS. Sec. 1753-A 1756 (1991)	Amend and add to certain provisions of geographic-based information services	This act establishes the Office of GIS in the Department of Conservation, which shall operate a geographic data base information center, develop and administer standards, subject to the approval of the Information Services Policy Board, and provide GIS services to the public, with licensing agreements.
May 22, 1989	Executive order establishing the Maine GIS Steering Committee	The committee is established to provide leadership, promote, plan, direct and coordinate statewide GIS and establish priorities, allocate governmental resources, and assure compatibility of information systems.
30-A M.R.S Sec. 4342 (1989)	Growth Management Program	Includes development of GIS
38 MRSA s 580-B, 546-B, Ch. 3	Oil Discharge Prevention and Pollution Control	Regarding the protection and improvement of waters and the discharge prevention and pollution control. Use of GIS to help determine sensitive area identification and protection.

GEOGRAPHIC INFORMATION/GIS COORDINATION GROUP

GIS Steering Committee (official by EO)
Mission: The committee's mission is to provide leadership and oversight in the development of a strategic plan and creation of a statewide GIS.
Sectors Represented: State, academic, utilities, regional entities, federal
Functions: Legislative, environmental protection-natural resource management, transportation, social services, public safety-emergency management, economic development

MARYLAND

AUTHORIZATION FOR GEOGRAPHIC INFORMATION/GIS

Citation	Name/Type	Description
Maryland Codes 10-902-905 (1992)	Automated mapping-GIS—access and fees	This statute states that governmental units may adopt a fee structure for products reflecting the cost of creating, developing, and reproducing the product. Services can be sold for a fee reflecting actual costs. It also provides that only people with a contract with a governmental entity can have on-line data access.

GEOGRAPHIC INFORMATION/GIS COORDINATION GROUPS

GIS Committee (unofficial)

Mission: The committee was organized to identify issues facing the development, implementation, use and maintenance of GIS in Maryland.
Sectors Represented: State, academic
Functions: -

Maryland State Government Geographic Information Coordinating Committee (official)

Mission: The purpose of MSGIC is to facilitate communication and cooperation among state agencies involved in the collection and use of spatial data and GIS, including minimizing duplication, supporting joint development of data, serving as a focal point and develop guidelines, and preserving valuable data.
Sectors Represented: State
Functions: V - governor's office, state planning, revenue-property tax, cultural resources-archeological-historical, archives-records management, environmental protection-natural resource management, education, budget-comptroller, health, agriculture, transportation, social services, insurance regulation, personnel-human development, public safety-emergency management, economic development, administrative, higher education, judicial

MASSACHUSETTS

AUTHORIZATION FOR GEOGRAPHIC INFORMATION/GIS

Citation	Name/Type	Description
GL Ch. 21A s.2 of the Acts of 1974	Authorizes the Executive Office of Environmental Affairs to have data responsibilities	This section authorizes the Executive Office of Environmental Affairs with responsibility for creating and collecting data, and acting as a clearinghouse for data to assist units of government and the private sector in making environmental decisions (this is the statute under which EOEA has GIS and distributes data).
GL 21A Ch240 Acts of 1989 sec2C 2001-1001	Data services and digital data fees	This section authorizes the Executive Office of Environmental Affairs to render data processing services and to distribute digital cartographic and other data according to fee schedule.

GEOGRAPHIC INFORMATION/GIS COORDINATION GROUP

Geographic Information Committee (official by Information Systems Office)

Mission: MGIC, formally organized by the Office of Management Information Systems, was established to coordinate geographic information activities in the state, including promoting collaboration, providing a forum for development of standards, technical assistance, and advising the public and private sectors.

Sectors Represented: State, local, regional entities, academic, private sector users

Functions: V - environmental protection-natural resource management, economic development, state planning; NV - census data center, information policy, information technology

MICHIGAN

AUTHORIZATION FOR GEOGRAPHIC INFORMATION/GIS

Citation	Name/Type	Description
Michigan Public Act 236, 1990	Additional recording fees	This act increases the recording fee of the first page of any legal instrument assessed by each county register of deeds. This fee was increased from $5.00 to $9.00, with $2.00 of these fees retained by counties. The other funds are deposited in the Survey and Remonumentation Fund.
Michigan Public Act 204, 1979	Michigan Resource Inventory Act	The act creates the Michigan Resource Inventory Program to provide for a land resource and current use inventory in the state, to provide for technical assistance and create an inventory advisory committee in and duties of the Department of Natural Resources, and to provide funding to localities for their participation.
Michigan Public Act 345, 1990	State Survey and Remonumentation Act	The duties of the commission include the preservation of all land survey records of vertical and horizontal monuments, and the encouragement of remonumentation programs in the state's counties. Each county shall establish a county monumentation and remonumentation plan, based on a model plan developed by the commission.

GEOGRAPHIC INFORMATION/GIS COORDINATION GROUPS

GIS Committee (official)
Mission: The role of the group is to improve coordination of GIS activities to meet the needs of the Department of Natural Resources.
Sectors Represented: State
Functions: Environmental protection-natural resource management

IMAGIN Consortium (official)

Mission: The IMAGIN (Improving Michigan Access to Geographic Information Network) Consortium was organized to respond to the need for digital geographic data, training, and coordination, with the goal of improving the access, distribution, updating, and public awareness of available data.

Sectors Represented: State, local, tribal, nonprofit organization, federal, regional entities, academic

Functions: Core members - environmental protection-natural resource management, legislative, library; others - governor's office, budget-comptroller, information policy, revenue-property tax, administrative, census data center, state surveyor, public lands management, forestry, state geological survey, cultural resources-archeological-historical, agriculture, transportation, health, community-local affairs, public safety-emergency management

Land Information Exchange (MLINK) (unofficial)

Mission: MLINK is a nonprofit organization dedicated to the development and use of automated land information systems maintained by public and private organizations in the state. It works to develop standards, specifications, and organizational structure to create a uniform system.

Sectors Represented: Local, utilities, regional entities, private sector users, private sector suppliers, local government via association

Functions: Environmental protection-natural resource management

MINNESOTA

AUTHORIZATION FOR GEOGRAPHIC INFORMATION/GIS

Citation	Name/Type	Description
ML 1993 Ch. 172 Sec. 14, subd. 8(a)	Appropriation of funds as recommended by the Leg. Com. on Minn. Resources - Base Maps	This appropriation is to provide the state share of a 50/50 match program with USGS to continue statewide coverage of orthophoto maps and update mapping for the state's major urban areas, and plan for future cooperative mapping and air photo programs.
ML 1993 Ch. 172 Sect. 14 subd. 14	Data Collected by LCMR recommended projects must be compatible with state spatial data	Data collected by projects recommended by the Legislative Commission on Minnesota Resources (LCMR) that have common value for natural resource planning and management must conform to information architecture defined by standards adopted by the State Information Policy Office. Data must integrate with LMIC data.
MN Statutes 1992 Ch. 16B.92	Enabling legislation for the Land Management Information Center (LMIC)	This describes LMIC's purpose is to foster the integration of environmental data, and provide services in computer mapping and graphics, environmental analysis, and small systems development. It also states when fees may be assessed for the center's products and services.
Executive Order 91-16 Sept. 16, 1991	Establishment of a Governor's Council on Geographic Information	The executive order provides that the council was established to promote efficient and effective use of resources by leading the development, management and use of GI in the state. The council makes recommendations in areas including policies, institutional arrangements, standards, education, stewardship, and others.
MN Statutes 1991 Sup. 103H.175 subd.2	Establishment of a Groundwater Monitoring Data Base	The Land Management Information Center shall maintain a computerized data base of the results of groundwater quality monitoring and make it accessible to the principal state environmental agencies.

ML 1993 Ch. 172 Sect. 14 subd. 8(b)	Rural county use of National Aerial Photography Program flight data	This appropriation is for a contract with Houston County to evaluate the quality of digital planimetric map products created from NAPP in effectively meeting county data needs, and to assist other counties in the future use of these products.
S.F. No. 1620, Ch.192 Sect. 74 (1993)	Transfer of Land Management Information Center	This legislation transferred the Land Management Information Center from the Department of Administration to the Office of Strategic and Long-Range Planning effective July 1, 1993.

GEOGRAPHIC INFORMATION/GIS COORDINATION GROUPS

Governor's Council on Geographic Information (official by EO)

Mission: The council promotes the efficient and effective use of resources by providing leadership and direction in the development, management, and use of GI in Minnesota. The Council makes recommendations in areas including, but not limited to, policies, standards, education, and stewardship.

Sectors Represented: State, local, federal, academic, utilities

Functions: Governor's office, state planning, information policy, legislative, administrative, environmental protection-natural resource management, transportation, higher education

GIS/LIS Consortium (unofficial)

Mission: The mission of the consortium is to provide a forum and serve as a users group for communicating and sharing information about GIS and land information systems in the state and to support the Governor's Council on Geographic Information.

Sectors Represented: State, regional entities, local, academic

Functions: Administrative, information technology, environmental protection-natural resource management, transportation, state planning, higher education

Department of Transportation Council for GI (official)

Mission: The council is managed by the Department's Office of Information Policy with the mission of increasing access to GI, increasing its accuracy, and increasing its ability to be integrated. These efforts will benefit those who develop and deliver the department's products and services.

Sectors Represented: State

Functions: State planning, information policy, transportation, higher education, public safety-emergency man,.. information technology

MISSISSIPPI

AUTHORIZATION FOR GEOGRAPHIC INFORMATION/GIS

Citation	Name/Type	Description
MRS 57-13-23, 1986	Establishes the Mississippi Automated Resource Information System	The MARIS' purpose includes storing, processing, and disseminating the state's natural and cultural resources consistent throughout state departments and, to the extent possible, with federal and privately generated resource data banks. The legislation established the MARIS Policy Committee, Task Force, and an Executive Committee.
Miss. Code Ann. 25-58-1,3 (1990)	Local Government GIS - authorization for borrowing	The board of supervisors of any county and the governing authority of any municipality are authorized and empowered to borrow money to create GIS and prepare multipurpose cadastre, with approval of Central Data Processing Authority.

GEOGRAPHIC INFORMATION/GIS COORDINATION GROUP

Automated Resource Information System Policy Committee (official via statute)

Mission: The mission of the Policy Committee is to direct the activities of MARIS, which was statutorily directed to include storing, processing, extracting, and disseminating data related to the state's natural and cultural resources.

Sectors Represented: V - state; **other** - academic, local, tribal, nonprofit organization, regional entities, private sector users

Functions: Information policy, information technology, agriculture, economic development, archives-records management, environmental protection-natural resource management, transportation, health, secretary of state, regulatory utilities commission, public safety-emergency management, revenue-property tax, state planning, governor's office, administrative, census data center, legislative, information policy, state geological survey, forestry, cultural resources-archeological-historical, economic development, higher education

MISSOURI

AUTHORIZATION FOR GEOGRAPHIC INFORMATION/GIS

None

GEOGRAPHIC INFORMATION/GIS COORDINATION GROUPS

Bio Diversity Council of Missouri (official)
Mission: The role of the council is to facilitate natural resource conservation planning by serving as a multiagency working group. One of its primary objectives is to encourage use of GIS and related technologies to meet this goal.
Sectors Represented: State, federal, academic
Functions: Environmental protection-natural resource management, public lands management, forestry, state geological survey, transportation, higher education

Systems Access & Information Consortium (MOSAIC) (official)
Mission: MOSAIC serves as the primary entity for promoting, coordinating, and supporting GIS applications, products, and data sharing in the state.
Sectors Represented: State, federal, local, academic
Functions: Environmental protection-natural resource management, public lands management, forestry, state geological survey, cultural resources-archeological-historical, agriculture, transportation, health, public safety-emergency management, community-local affairs, information technology, information policy

MONTANA

AUTHORIZATION FOR GEOGRAPHIC INFORMATION/GIS

Citation	Name/Type	Description
May 1990	Memorandum concerning interagency coordination and support for GIS in Montana	This MOU was signed by eleven federal entities, seven state agencies, and two universities with the purpose of establishing a vehicle for participating agencies and organizations to develop GIS in Montana. The MOU establishes the GIS Interagency Management Steering Committee and the GIS Interagency Technology Workgroup.
Code Annotated 90-15 101, 1985	Natural Resource Information System	The State Library was directed to plan for and implement the system with recommendations provided by a Natural Resource Data System Advisory Committee established in the statute. The committee's role includes making recommendations regarding criteria for data types and categories, collection format, and identity of data sources.

GEOGRAPHIC INFORMATION/GIS COORDINATION GROUPS

GIS Interagency Technical Working Group (official by MOU)
Mission: The group strives to facilitate joint projects, inventory digital data, create a data documentation format, develop and compile data standards for the common base themes, investigate methods and priorities towards creation of a statewide transferable digital data base, and disseminate the data.
Sectors Represented: State, federal, tribal, academic
Functions: Library, environmental protection-natural resource management, public lands management, transportation

GIS Users Group (unofficial)
Mission: The users group is a consortium of governmental agencies and private businesses involved with GIS technology. The group provides a forum for exchanging information and ideas.
Sectors Represented: State, federal, local, private sector users, private sector suppliers
Functions: Library, environmental protection-natural resource management, public lands management, transportation

Interagency GIS Management Steering Committee (official by MOU)
Mission: The purpose of the committee is to set statewide policies, and create an action plan to implement the objectives of the MOU while guiding the long-term direction of GIS in the state and the activities of the Technical Working Group. The group also ensures these agencies comply with data standards.
Sectors Represented: State, federal, tribal, academic
Functions: Library, environmental protection-natural resource management, public lands management, transportation

NEBRASKA

AUTHORIZATION FOR GEOGRAPHIC INFORMATION/GIS

Citation	Name/Type	Description
LB 541 (Approp. Bill) Sec.16, no.18, 1993)	Provide direction to the State GIS Steering Committee	This bill expresses the intent of the legislature that the GIS Steering Committee will carry out specific duties including, develop a comprehensive inventory of state agency uses of GIS, prepare a survey regarding future uses, develop a process to meet future needs, and develop recommendations for standards and help coordination.
Neb.Rev.St. s 81-2601-2605 (1991)	To create the GIS Steering Committee	The committee's charge is to facilitate acquisition, compatibility, and communications of GIS technology at all levels of government, make recommendations to the legislature, and establish guidelines and policies for statewide GIS, including quality assurance and control, enforcement mechanisms, access, cost recovery, and priorities.

GEOGRAPHIC INFORMATION/GIS COORDINATION GROUP

GIS Steering Committee (official via statute)

Mission: The committee's role is to facilitate acquisition, compatibility and communication of GIS technology at all levels of government, make recommendations for program initiatives and funding, establish guidelines and policies for quality control, access, cost recovery, ownership, etc.

Sectors Represented: State, utilities, regional entities, academic, local government via association

Functions: Administrative, governor's office, environmental protection-natural resource management, state surveyor, legislative, state planning, transportation, public lands management, state geological survey

NEVADA

AUTHORIZATION FOR GEOGRAPHIC INFORMATION/GIS

Citation	Name/Type	Description
NRS 218.051, 17, 218; NRS 304.060, 24, 304	Congressional Districts - use of GIS in redistricting	These two statutes (1991) authorized and provided funding, respectively, to use GIS for reapportionment of legislators and districts, as well as for Senate and House elections districting.

GEOGRAPHIC INFORMATION/GIS COORDINATION GROUP

State Mapping Advisory Committee (official by EO)

<u>Mission</u>: The SMAC coordinates mapping needs and develops recommendations for mapping priorities for federal agencies with mapping activities, resulting in various cooperative projects. The role was expanded to include digital data and GIS in the late 1980s, including forming a GIS Subcommittee.

<u>Sectors Represented</u>: **V** - state; **NV** - academic, federal, local

<u>Functions</u>: Legislative, geological survey, information policy, information technology, environmental protection-natural resource management, public lands management, transportation, library, public safety-emergency management

NEW HAMPSHIRE

AUTHORIZATION FOR GEOGRAPHIC INFORMATION/GIS

Citation	Name/Type	Description
RSA Ch. 102 (HB 899) (1988)	Computer assistance to regional planning commissions	This act provided for an appropriation of $270,000 for nine regional planning agencies of the state to have "computer interface capability" with each other, the Office of State Planning (OSP), and state data collection and storage sources including compatibility with OSP's GIS and other similar data sources.
RSA Ch. 4-C:8, 1989	Regional and municipal assistance	It states that the Office of State Planning shall provide technical assistance to municipalities in the (a) use and application of data in the state's GIS for local planning and growth management, and (b) recommending standards for large-scale mapping for municipal functions such as tax assessment and public facility management.
RSA Ch. 198 (1993)	Requiring the Office of State Planning to conduct a satellite survey of clearcut areas	This act requires the Office of State Planning, in cooperation with the Division of Forest and Lands, to study clearcut information and contract with the University of New Hampshire for purchase and analysis of satellite imagery to survey clearcutting in the state. Funds were appropriated, including specifically for GIS.

GEOGRAPHIC INFORMATION/GIS COORDINATION GROUP

GIS Advisory Committee of the Council on Resources and Development (official by statute)

Mission: CORD's role is to consult on common problems in environmental protection, natural resources and growth management, to coordinate and resolve differences, and to make recommendations. It created the GIS committee to coordinate and recommend data base strategies, standards, and financing among governments.

Sectors Represented: State

Functions: State planning, environmental protection-natural resource management, economic development, transportation, agriculture, public safety-emergency management, health, education

NEW JERSEY

AUTHORIZATION FOR GEOGRAPHIC INFORMATION/GIS

Citation	Name/Type	Description
Public Law Ch. 78, 1990	An act concerning hazardous substance discharge prevention	Oil companies are required to map their facilities, pipelines, and off-site land and water areas, which could be adversely affected by a discharge. The Department of Environmental Protection is to develop maps of wetlands, shellfish, waterways, and coastal areas to provide coverage of the rest of the state.

GEOGRAPHIC INFORMATION/GIS COORDINATION GROUP

State Mapping Advisory Committee (official by EO)

Mission: SMAC serves as a coordination mechanism and forum for exchanging ideas and discussing mapping issues, and to consolidate statewide mapping requirements to USGS. Focus was expanded in 1990 to include GIS, including policies for statewide approach and subcommittees to address such issues.

Sectors Represented: State, academic, professional organization, utilities, local, private sector suppliers, private sector users, regional entities, federal

Functions: Environmental protection-natural resource management, community-local affairs, transportation, revenue-property tax, employment-labor, agriculture, public safety-emergency management, state planning, census data center, state geological survey

NEW MEXICO

AUTHORIZATION FOR GEOGRAPHIC INFORMATION/GIS

Citation	Name/Type	Description
Executive Order 1987	Authorizing the New Mexico Geographic Information Council	This executive order authorizes the council, which was initially organized in 1984. It provides that the council provide recommendations regarding geographic data needs, priorities, and standards to the governor, and to state, federal, and local agencies and also provide recommendations to the U.S. Board on Geographic Names.
The Research and Development Act, 1988	Providing for development, implementation, and maintenance of a resource GIS	This statute provided for the original feasibility study for the Resource GIS Program managed by the University of New Mexico and coordinated with state government. The statute provides that RGIS include provision of a "comprehensive, state-of-the-art, automated GI clearinghouse" as a management, planning, and analysis tool.

GEOGRAPHIC INFORMATION/GIS COORDINATION GROUPS

GIS Advisory Committee of the Information Systems Council (official, ISC by statute)

<u>Mission</u>: GISAC's role as defined by ISC includes responsibility for making GIS policy to ISC, for establishing GIS standards, and conducting GIS procurements on behalf of the Council. Agencies are directed to conform to such standards.

<u>Sectors Represented</u>: State, local government via association, academic, federal, private sector-users

<u>Functions</u>: Administrative, information technology, environmental protection-natural resource management, revenue-property tax, census data center, public lands management, cultural resources-archeological-historical, public safety-emergency management, transportation, archives-records management, personnel-human development, forestry, health, regulatory utilities commission, higher education, economic development, community-local affairs, state geological survey

Geographic Information Council, Inc. (official by EO)
Mission: NMGIC goals are to promote a comprehensive approach to geographic information, foster coordination of programs, policies, technologies, and resources to maximize opportunities and minimize duplication, and make recommendations for all governments about data needs, priorities, and standards.
Sectors Represented: State, federal, local, academic, private sector users, private sector suppliers, tribal, professional organization
Functions: Environmental protection-natural resource management, transportation

NEW YORK

AUTHORIZATION FOR GEOGRAPHIC INFORMATION/GIS

Citation	Name/Type	Description
Senate Bill 6087 (1991)	Authorizing Erie Co. Water Authority to sign contracts for GIS and mapping, and sell data, services	
McKinney's Laws - NY 43-B, 44-0117	Hudson River Valley GIS	Authorized GIS as part of Hudson River Valley Greenway effort.

GEOGRAPHIC INFORMATION/GIS COORDINATION GROUP

State Forum for Information Resource Management (official by budget director)

Mission: The IRM Forum was created to increase coordination of information resources and technologies, through serving as a network, clearinghouse, and educational resource. It has recently focused on GIS as a platform to discuss data sharing and institutional issues, including demonstration projects.

Sectors Represented: **V** - state, academic; **NV** - local, nonprofit organization

Functions: Budget-comptroller, archives-records management, personnel-human development, environmental protection-natural resource management, transportation, health, legislative, library, revenue-property tax, secretary of state, insurance regulation, regulatory utilities commission, social services, employment-labor, education, higher education, public safety-emergency management, economic development, judicial, census data center

NORTH CAROLINA

AUTHORIZATION FOR GEOGRAPHIC INFORMATION/GIS

Citation	Name/Type	Description
Executive Order July 30, 1991	Establishes the GI Council and transfer the Center for GI and Analysis to Governor's Office	Roles and responsibilities of these identities are defined. Identifies the Center for Geographic Information and Analysis as the lead GIS agency in state government.
1991 H.B. 356	Exception to the Public Records Act for GIS in Lincoln and Brunswick counties	Provides that electronic copies of geographic data provided at reasonable cost to the public may not be resold or redistributed for trade or commercial products. This direction applies to the Counties of Guilford, Pitt, Mecklenburg, and Nash; and for the Cities of Greensboro and High Point.
1993 H.B. 143, Ch. 82	Exception to the Public Records Act for GIS in specified counties and cities	
Executive Order May 21, 1993	Expands composition of GI Council and establishes Info. Resources Management Comm. as oversight	This executive order reestablishes the Geographic Information Coordinating Council in a new administration, including its reporting to the Information Resources Management Commission, and expansion of its membership to include federal, local, and regional representation as well as increased state agency participation.
Gen. Stat. 143-345.6 1977	Land Records Management Program and creation of Advisory Committee	This statute directs a statewide land records management program to provide technical and financial assistance to counties to modernize local land records and minimum standards for indexing of land records, maps, and security and reproduction of land records. It also provided for a common parcel ID numbering scheme statewide.
Memorandum of Agreement, August 1992	Establishes statewide network of three base stations for GPS	This MOA establishes a cooperative effort between the Center for GI and Analysis, State Information Processing Services, Department of Transportation, and the Department of Environment, Health and Natural Resources to install and maintain GPS Base Station Network useable by all.

1991 N.C. Ch. 285	Qualified exception from Public Records Act for certain GIS	County may require agreement that GIS data not be resold for commercial purposes.
Senate Bill 583 (1991)	Reuse of data in Catawba County GIS	Provides that if Catawba County has its GIS data base and data, and makes electronic and hard copy access at reasonable cost to public; person receiving electronic copy may not resell that information or use it for commercial purposes.

GEOGRAPHIC INFORMATION/GIS COORDINATION GROUP

Geographic Information Coordinating Council (official by EO)

<u>Mission:</u> The role of the council is to foster cooperation among all governments, academia, and the private sector regarding GI; improve the quality, access, cost effectiveness, and utility of GI; promote GI use and consideration as a strategic resource, be a coordination framework, and share resources.

<u>Sectors Represented:</u> State, local, professional organization, federal, regional entities, private sector users

<u>Functions:</u> V - state planning, budget-comptroller, administrative, environmental protection-natural resource management, agriculture, health, transportation, secretary of state, education; NV - information policy

NORTH DAKOTA

AUTHORIZATION FOR GEOGRAPHIC INFORMATION/GIS

None

GEOGRAPHIC INFORMATION/GIS COORDINATION GROUPS

GIS Management Steering Committee (official by budget director)
Mission: The role of the steering committee is to provide direction and coordination for GIS.
Sectors Represented: State, academic, federal
Functions: Budget-comptroller, environmental protection-natural resource management, cultural resources-archeological-historical, census data center, public lands management

GIS Technical Advisory Committee (official by Management Steering Committee)
Mission: The role of the Technical Advisory Committee is to concentrate on technical issues and recommend standards for adoption by the Steering Committee, as well as develop a data inventory, inventory of statewide interests and needs, provide education and recommend a standard base map.
Sectors Represented: State, academic, federal, local government via association
Functions: Environmental protection-natural resource management, health, transportation, regulatory utilities commission, information technology, economic development, public lands management, agriculture, state geological survey

OHIO

AUTHORIZATION FOR GEOGRAPHIC INFORMATION/GIS

Citation	Name/Type	Description
Executive Order 93-010-V	Authorizes Ohio Geographic Referenced Information Program and Council and Forum	This order establishes the Ohio Geographic Referenced Information Program (OGRIP) composed of a council and a forum. It provides that the council's role is to coordinate GIS in the state providing for the efficient collection, management, and use of GI, and establish a GIS Forum to assist activities and encourage access and consistency.

GEOGRAPHIC INFORMATION/GIS COORDINATION GROUPS

Geographic Referenced Information Program (OGRIP) Council (official by EO)

Mission: The OGRIP Council's mission includes encouraging creation of multiple use data, facilitating determination and access to available data and encouraging its use, and generally coordinating GI/GIS activities in the state to provide for efficient collection, management, and use of data.

Sectors Represented: State, academic, local, utilities

Functions: Information policy, administrative, information technology, environmental protection-natural resource management, transportation, higher education, community-local affairs

OGRIP Forum (official by EO)

Mission: The mission of the OGRIP Forum is "to assist in the coordination of GIS activities and to encourage access and consistency with other GIS systems to the maximum extent possible."

Sectors Represented: State, local, local government via association, federal, academic, utilities

Functions: Budget-comptroller, secretary of state, environmental protection-natural resource management, transportation, regulatory utilities commission, employment-labor, higher education, public safety-emergency management, economic development, administrative, information technology, census data center

OKLAHOMA

AUTHORIZATION FOR GEOGRAPHIC INFORMATION/GIS

None

GEOGRAPHIC INFORMATION/GIS COORDINATION GROUP

GIS Technology Policy Board (official by Governor)

Mission: The board was established to encourage the development and implementation of GIS while coordinating state activities to promote the sharing and integration of information. It is developing policies regarding standards, data base development and mapping, access, and funding.

Sectors Represented: State

Functions: Environmental protection-natural resource management, transportation, economic development, agriculture, budget-comptroller, education, health, social services, public safety-emergency management, legislative, census data center, public lands management, state geological survey, regulatory utilities commission, higher education

OREGON

AUTHORIZATION FOR GEOGRAPHIC INFORMATION/GIS

Citation	Name/Type	Description
Executive Order 89-16, Oct. 10, 1989	Establishing State Map Advisory Council	This order affirmed previous orders establishing SMAC's functions including planning, policy and technical issues, technical assistance and coordination, developing a state GI data base, determining data custodianship roles of agencies, and managing some data. This work to be supported by State GIS Center also created by order.
ORS 190.050, 192.502 (1991)	Intergovernmental cooperation - fees for geographic data and uses	This act provides that notwithstanding any other provisions of law, an intergovernmental group's geographic data bases or systems are confidential and exempt from public disclosure. Access cannot be restricted, but fees can be charged based on market prices or competitive bids for geographic data that have commercial value.
ORS s 196.575 (1989)	Ocean resources management - data liaisons and use	Authorizes obtaining oceanographic data from the federal government and joint liaison program. This includes requirement to develop GIS data base for program.

GEOGRAPHIC INFORMATION/GIS COORDINATION GROUP

State Map Advisory Council (official by EO)

Mission: SMAC's directed purposes are strategic planning, resolving policy and technical issues and disputes, providing technical assistance and coordination, standardization of data acquisition, review budget requests for GI/GIS, and make a statewide GI data base and plan to accomplish it.

Sectors Represented: V - state; NV - federal, local, regional entities, academic, local government via association

Functions: Information policy, information technology, environmental protection-natural resource management, transportation, public lands management, revenue-property tax, social services, census data center, forestry, state geological survey

PENNSYLVANIA

AUTHORIZATION FOR GEOGRAPHIC INFORMATION/GIS

None

GEOGRAPHIC INFORMATION/GIS COORDINATION GROUP

GIS Subcommittee of Automated Technology Steering Committee (official by governor's assistant)

Mission: The GIS Subcommittee's mission is to formulate appropriate GIS implementation methodologies, foster cooperation between state agencies, eliminate duplication in planning and implementing GIS, enhancing data sharing through GIS, and develop a planning methodology for GIS evaluation.

Sectors Represented: State

Functions: Library, census data center, information policy, environmental protection-natural resource management, transportation, agriculture, economic development, community-local affairs, health, administrative, archives-records management, forestry, state geological survey, cultural resources-archeological-historical, public lands management, public safety-emergency management, employment-labor

RHODE ISLAND

AUTHORIZATION FOR GEOGRAPHIC INFORMATION/GIS

Citation	Name/Type	Description
GL Ch 42-11 Sec 2,10 Amd Ch 16-32, 1990	Department of Admin. and University of Rhode Island Amendment	This act officially established the Rhode Island GIS (RIGIS) and directs the Dept. of Administration's Division of Planning to manage it. It also directs the state planning council establish an executive committee consisting of major participants of RIGIS. The University of Rhode Island was authorized to provide technical support and assistance.
Gen.Laws s 46-15.4-3 1991	Water Supply Management Plans - Content	This statute requires that Water Supply Management Plans must have certain components, including specified data requirements.

GEOGRAPHIC INFORMATION/GIS COORDINATION GROUP

RIGIS Executive Committee (official via statute)

Mission: The committee serves as a policy level coordination group that adopts policies and standards for the Rhode Island GIS, as well as establishing its direction, priorities, and data layers. Goals are accurate, organized and documented GIS, avoid duplication, and use GIS for land management.

Sectors Represented: State, federal, local, academic

Functions: State planning, environmental protection-natural resource management, transportation

SOUTH CAROLINA

AUTHORIZATION FOR GEOGRAPHIC INFORMATION/GIS

None

GEOGRAPHIC INFORMATION/GIS COORDINATION GROUP

State Mapping Advisory Committee (official by EO)
Mission: The SMAC's main purpose has been to act as an advisory and educational group, providing networking among participants. It evolved to include property tax mapping, geodetic control, and GIS.
Sectors Represented: V - state, local, utilities, academic, local government via association, professional organization, private sector users; NV - federal
Functions: Information policy, information statistics, environmental protection-natural resource management, economic development, health, revenue-property tax, transportation, regulatory utilities commission

SOUTH DAKOTA

AUTHORIZATION FOR GEOGRAPHIC INFORMATION/GIS

None

GEOGRAPHIC INFORMATION/GIS COORDINATION GROUP

GIS Advisory Committee (official via governor's support)

Mission: The role of the committee is to facilitate coordination and minimize redundancy, including conducting needs assessments and setting statewide priorities. It reviews technical options considering connectivity needs and sponsors projects and educational and networking events.

Sectors Represented: State

Functions: Budget-comptroller, information policy, information technology, environmental protection-natural resource management, agriculture, education, health, revenue-property tax, public lands management, transportation

TENNESSEE

AUTHORIZATION FOR GEOGRAPHIC INFORMATION/GIS

Citation	Name/Type	Description
Tennessee Code Ann. 10-7-506	Public records with commercial value - fees may be charged accordingly	This statute provides that if a request is made for a copy of a public record that has commercial value and it requires reproduction of a part of an automated map or other geographic data developed with public funds, the county may impose fees including development and maintenance costs.

GEOGRAPHIC INFORMATION/GIS COORDINATION GROUPS

GIS Steering Committee of the Information Systems Council (ISC official by EO)
Mission: The role of the committee is to help develop information standards, particularly those for cartography and GIS, while also encouraging coordinated development of GIS and avoiding duplication.
Sectors Represented: State, local government via association
Functions: Information policy, information technology, state planning, budget-comptroller, environmental protection-natural resource management, revenue-property tax, transportation, health, employment-labor, education, higher education

GIS Working Group (official by Steering Committee)
Mission: The mission of the GIS Working Group is to analyze the status of GIS in the state, primarily within state government, and also to provide networking and information exchange among GIS users.
Sectors Represented: State, federal, local, academic, private sector users, private sector suppliers
Functions: Environmental protection-natural resource management, state planning, transportation, health

TEXAS

AUTHORIZATION FOR GEOGRAPHIC INFORMATION/GIS

Citation	Name/Type	Description
Executive Order AWR 92-6, 1992	Providing direction to the GIS Planning Council	This order provides that the Department of Information Resources has chartered the GIS Planning Council and directs that the Council shall plan for the most effective means of acquiring and distributing GI and ensuring agencies are in consort with state and federal agencies, develop a business plan, and ID data custodians and roles.

GEOGRAPHIC INFORMATION/GIS COORDINATION GROUPS

Geographic Information Systems Planning Council (official by EO)
Mission: The role of the council is to plan for the most cost-effective means of distributing GI in the state, and ensure agency programs are in concert with other state and federal agencies. Agencies are to be identified, and roles and responsibilities defined for data custodians.
Sectors Represented: State, local government via association, regional entities, academic, private sector via association
Functions: Information policy, environmental protection-natural resource management, public lands management, transportation, health, social services, education, public safety-emergency management, community-local affairs, governor's office, budget-comptroller, secretary of state, regulatory utilities commission, Railroad Commission

Texas Mapping Advisory Committee (historical)
Mission: The role of TMAC is to facilitate information exchange among agencies and persons working with mapping in Texas, and has provided input to USGS regarding mapping priorities. It expanded its scope to digital cartography in the mid-1980s and has a subcommittee accordingly.
Sectors Represented: State, regional entities, professional organization
Functions: Environmental protection-natural resource management

UTAH

AUTHORIZATION FOR GEOGRAPHIC INFORMATION/GIS

Citation	Name/Type	Description
Utah Code 63-1-61,62 1991	Automated Geographic Ref. Center (AGRC) & State Geographic Information Database (SGID)	This act creates SGID and mandates state agencies to comply with policies and standards. It authorizes AGRC, in the Department of Administrative Services, to provide GIS services and manage SGID, with standard format, lineage, etc. SGID is the central reference for all information in any GIS database, and a repository for data with multiple users.
House Con. Res. 24 1991	Utah Geographic Information Council	This resolution recognizes the council has been established and states the "Legislature and Governor commend and support" the efforts of the council.

GEOGRAPHIC INFORMATION/GIS COORDINATION GROUPS

Geographic Information Council (official via governor and legislature)
Mission: UGIC's purpose is to serve as an umbrella organization with a forum for communication and information exchange about manual and automated GI. It was designed to work with other groups to reduce overlap and redundancy between groups working with GI/GIS.
Sectors Represented: State, federal, local, local government via association, professional organization, private sector users, private sector -suppliers, academic, utilities
Functions: Environmental protection-natural resource management, cultural resources-archeological-historical

State GIS Advisory Committee of the Information Tech. Review Committee (ITRC via statute)
Mission: The role of the GIS Committee is to formulate and recommend proposed GIS policies, procedures, and standards; recommend priorities for data collection; review legal and policy issues related to data access; and make recommendations to resolve issues. It also oversees the state GI data base.
Sectors Represented: State, local, federal, academic
Functions: Information policy, information technology, environmental protection-natural resource management, transportation, agriculture, legislative

VERMONT

AUTHORIZATION FOR GEOGRAPHIC INFORMATION/GIS

Citation	Name/Type	Description
1992 H. B. 955	Authorizes governor to make Vermont GIS a not-for-profit corporation	This bill amends the 1988 bill that originally authorized GIS in Vermont. It provides that the strategy for GIS may include creation of a not-for-profit corporation to serve governments but not compete with private sector services. It establishes a Board of Directors, specifies its members and defines some roles.
24 VSA 4303, 4306, 4325,4345a Revised 1988	Develop regional data base compatible with GIS, and establish duties of planning commissions	Establishes powers and duties of regional planning commissions, including development of a regional data base compatible with GIS.
Executive Order 92-3A	Creates the Vermont Center for GI, Inc. as lead for state GI coordination	The order creates the Vermont Center for GI, Inc. to foster GIS development and use. It establishes the corporate structure including the membership of the Board of Directors, powers of the corporation including data access procedures, standards, etc. The order also directs financial and planning requirements.
VSA Title 3, Sec 20 1988	The Growth Management Act relating to encouraging local, regional, and state agency planning	The act directed the governor to "develop a comprehensive strategy for the development and use of a GIS" and appropriated $4.75 million over five years for GIS. It provides the governor and regional planning districts shall ensure that all data relevant for GIS are developed in a compatible way for GIS use.

GEOGRAPHIC INFORMATION/GIS COORDINATION GROUPS

Vermont Center for GI, Inc. Board of Directors (official by statute)
Mission: The mission of the VCGI Board is direct and manage the VCGI, including developing, publishing, maintaining and implementing policies, procedures and standards and providing access to the Vermont GIS. This role also includes providing GI services, products and support to public and private users.
Sectors Represented: State, local government via association, regional entities, academic
Functions: Administrative, legislative, environmental protection-natural resource management, transportation, economic development, community-local affairs

VGIS Technical Advisory Committee
Mission: The mission of the Technical Advisory Committee is to work with and support the Vermont Center for GI, including providing input and advise from state agency and other users. It also serves as a GIS users group providing information exchange among members.
Sectors Represented: State, local, professional organization, regional entities, academic, private sector suppliers, private sector users
Functions: Environmental protection-natural resource management, transportation

Mapping Advisory Committee
Mission: The purpose of the committee is to provide recommendations regarding mapping priorities, including to USGS.
Sectors Represented: State, regional entities, professional organization, academic
Functions: Environmental protection-natural resource management, public lands management, forestry, state geological survey, transportation

VIRGINIA

AUTHORIZATION FOR GEOGRAPHIC INFORMATION/GIS

Citation	Name/Type	Description
Code of Virginia Act Ch. 668, 1989	Council on Information Management to conduct a study of land use information or mapping systems	CIM was directed to assess the need and utility of geographic and biological land use information or mapping systems, identify opportunities for system integration and shared usage, make recommendations, and develop a plan for the coordinated operation and development of such systems.
Code of Virginia, 2.1-563.32, 1992	Created the Advisory Committee on Mapping, Surveying, and LIS to advise Council on Information Management	Created a new Advisory Committee (replacing the former Commission) on Mapping, Surveying, and LIS. The committee meets with, confers with, and advises the Council on Information Management on matters relating to mapping, surveying and land information systems.
Code of Virginia, 15.1-11.7 1992	Local Government GIS	Provides that local governments may develop GIS and require their departments to use them.

GEOGRAPHIC INFORMATION/GIS COORDINATION GROUP

Advisory Committee on Mapping, Surveying, and Land Information Systems (official via statute)

Mission: The role of the Committee is to meet with, confer with and advise the Council on Information Management regarding mapping, surveying, and LIS, particularly for development of policies, standards, procedures, guidelines, recommendations, and access to data, while encouraging coordination.

Sectors Represented: State, local, professional organization, regional entities, utilities

Functions: Transportation, public safety-emergency management

WASHINGTON

AUTHORIZATION FOR GEOGRAPHIC INFORMATION/GIS

Citation	Name/Type	Description
Ch.58.22.020,24.020, 79.68.120, 1973,amended	Authorizes role of Dept. of Natural Resources in geographic information	Provides that (1) DNR shall establish and maintain a state base mapping system, and also define standards for it, (2) DNR's division of engineering is the official agency for surveys and maps, and (3) DNR shall design expansion of its land use data bank to include information to assist in land use, growth, and influence environmental quality.
RCW 43.63A.550, 1990	Growth Management Act	The act requires specific cities and counties to engage in planning activities, including comprehensive plans. The act requires collecting data on public and private land for land use, demographics, infrastructure, critical areas, housing, etc. It directs an advisory group with planning and GIS expertise.

GEOGRAPHIC INFORMATION/GIS COORDINATION GROUP

State Geographic Information Council (official via Department of Information Systems)

Mission: WSGIC was formed by the Department of Information Services to serve as a management level forum for exploring issues and alternatives, exchanging information, developing and recommended standards, and promoting cooperative data efforts. It fosters GI coordination and provides overall GI/GIS direction.

Sectors Represented: V - state, academic, federal, local, private sector users; **NV** - tribal, nonprofit organization

Functions: Information policy, information technology, environmental protection-natural resource management, health, economic development, community-local affairs, public safety-emergency management, transportation, employment-labor, social services, education, public lands management, budget-comptroller, library, legislative, forestry

WEST VIRGINIA

AUTHORIZATION FOR GEOGRAPHIC INFORMATION/GIS

None

GEOGRAPHIC INFORMATION/GIS COORDINATION GROUPS

GIS Coordinating Committee (unofficial)

Mission: The Committee was formed to provide a forum for GIS users and potential users to share ideas and information about technology, applications, and data exchange. It is working to determine a future direction for the state regarding GIS, including GPS as its initial effort.

Sectors Represented: State, academic, federal, local, private sector suppliers, private sector users, regional entities, local government via association

Functions: Environmental protection-natural resource management, health, transportation, state planning, state geological survey, higher education, economic development

State GIS Steering Committee (unofficial, pending official status)

Mission: The committee is being officially organized, with a pending executive order, and based on the existing coordination group. The role will be to encourage coordination and deal with technical issues to further GIS activities.

Sectors Represented: State, academic, federal, local, private sector suppliers, private sector users, local government via association, regional entities, utilities

Functions: Environmental protection-natural resource management, health, transportation, state planning, governor's office, higher education, economic development, community-local affairs, education, census data center, administrative, public safety-emergency management, regulatory utilities commission, public lands management, information statistics, cultural resources-archeological-historical

WISCONSIN

AUTHORIZATION FOR GEOGRAPHIC INFORMATION/GIS

Citation	Name/Type	Description
Wisconsin Laws 31 (1989)	Creating the Land Information Program and the Land Information Board	The board mission was established to direct and supervise the program, including serving as the state clearinghouse for land information and providing technical assistance to state and local agencies. Aid to counties is included, including designation of a county land information office and the opportunity to receive grants if participating.
Wisconsin Laws 339 (1990)	Establishing funding for the Land Information Program and ending matching fund needs to receive grants	It provides recording fees be increased for first pages of legal documents from $4 to 8 in July, 1990, and to $10 in July, 1991, with counties retaining $2 of the first $4, and all of the $2 added in 1991. Counties are eligible for grants for the other funds, but must have designated office and plan for land records modernization.

GEOGRAPHIC INFORMATION/GIS COORDINATION GROUPS

Land Information Association (unofficial)

Mission: WLIA's mission is to promote integrated and multipurpose LIS networks and policy through legislative activity, membership education, standards development, forums for emerging concepts, and outreach. Its four goals include policy, networking, technical, and education with efforts in each area.

Sectors Represented: State, federal, local, regional entities, academic, professional organization, utilities, private sector suppliers, private sector users, tribal, nonprofit organization

Functions: Information policy, revenue-property tax, insurance regulation, environmental protection-natural resource management, public lands management, state geological survey, agriculture, transportation, economic development, community-local affairs, rural development, legislative, information technology, census data center, higher education, state cartographer, regulatory utilities commission

Land Information Board (official via statute)
Mission: The Board directs the Land Information Program, with a major goal of systems integration to ensure multiple use and sharing of data by all. The Board develops guidelines, administers a local grant program, provides technical assistance to state and local agencies, and serves as a data clearinghouse.
Sectors Represented: V - state, local, professional organization, private sector suppliers, utilities; NV - federal, regional entities, local government via association, academic, private sector via association
Functions: Revenue-property tax, environmental protection-natural resource management, public lands management, forestry, state geological survey, cultural resources-archeological-historical, agriculture, transportation, higher education, legislative, administrative, census data center, state cartographer

State Inter-Departmental Geographic Data Sharing Workgroup (unofficial)
Mission: The workgroup's mission is to discuss, produce, implement and promote technical solutions for the processing of digital GI, with the goal to enhance the delivery of governmental services. The group solves day-to-day, operational problems, including collection, storage, and display of data.
Sectors Represented: State, local government via association, regional entities, utilities
Functions: Environmental protection-natural resource management, health, transportation, state geological survey, economic development, community-local affairs, education, state planning, administrative, higher education

WYOMING

AUTHORIZATION FOR GEOGRAPHIC INFORMATION/GIS

None

GEOGRAPHIC INFORMATION/GIS COORDINATION GROUPS

GIS Steering Committee (official by Computer Technology Division)
Mission: The Computer Technology Division of the Department of Admin. and Information established this group to set the direction of GIS efforts in the state. They set the agenda of the GIS Users Group as well as maintain the state metadata catalog, and address technical GIS issues.
Sectors Represented: state, local, federal, academic, private sector-suppliers, private sector-users
Functions: governor's office, library, revenue-property tax, secretary of state, legislative, administrative, information technology, census data center, archives-records management, environmental protection-natural resource management, public lands management, state geological survey, cultural resources-archeological-historical, transportation, social services, health, economic development

GIS Users Group (official by Computer Technology Division)
Mission: The Computer Technology Division of the Department of Admin. and Information created the group to promote communication, coordination, and sharing of data among various entities in the state, initially concentrating on state agencies. Effort is on data sharing and creation of a state base map.
Sectors Represented: state, local, federal, academic, private sector-suppliers, private sector-users
Functions: information technology, environmental protection-natural resource management, revenue-property tax, governor's office, library, secretary of state, legislative, administrative, census data center, archives-records management, public lands management, state geological survey, cultural resources-archeological-historical, transportation, social services, health, economic development

State Mapping Advisory Committee (official by state agency)
Mission: The role of the SMAC is coordinate activities among state agencies and others and provide input to USGS regarding mapping priorities.
Sectors Represented: state
Functions: environmental protection, natural resource management

ACRONYMS

AM/FM	Automated Mapping/Facilities Management (an association)
BLM	Bureau of Land Management
DLG	Digital Line Graph
DOQ	Digital Orthophoto Quarterquad
EPA	Environmental Protection Agency
FGDC	Federal Geographic Data Committee
FIPS	Federal Information Processing Standard
FWS	Fish and Wildlife Service
GCDB	Geographic Coordinate Data Base
GIS	Geographic Information Systems
GPS	Global Positioning System
IVHS	Intelligent Vehicle Highway System
MOU	Memorandum of Understanding
MSC	Mapping Science Committee
NASIRE	National Association of State Information Resource Executives
NII	National Information Infrastructure
NGS	National Geodetic Survey, NOAA
NMD	National Mapping Division, USGS
NOAA	National Oceanic and Atmospheric Administration
NSDI	National Spatial Data Infrastructure
NSGIC	National States Geographic Information Council
OMB	Office of Management and Budget
SCS	Soil Conservation Service
SCWRC	South Carolina Water Resources Commission
SDNR	State Department of Natural Resources
SDTS	Spatial Data Transfer Standard
SEPA	State Environmental Protection Agency
SIPC	State Information Policy Consortium
SMAC	State Mapping Advisory Committee
SSURGO	Soil Survey Geographic Data base
USGS	United States Geological Survey